DIGITAL
ELECTRONICS
TROUBLESHOOTING
2ND EDITION

No. 2750
$25.95

DIGITAL
ELECTRONICS
TROUBLESHOOTING

2ND EDITION
JOSEPH J. CARR

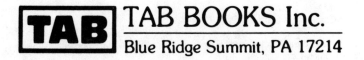 TAB BOOKS Inc.

Blue Ridge Summit, PA 17214

SECOND EDITION

FIRST PRINTING

Library of Congress Cataloging in Publication Data

Carr, Joseph J.
    Digital electronics troubleshooting.

    Includes index.
    1. Digital electronics.   2. Electronic apparatus and appliances—Maintenance and repair.   I. Title.
TK7868.D5C36   1987   621.3815   86-30168
ISBN 0-8306-0450-2
ISBN 0-8306-2750-2 (pbk.)

Questions regarding the content of this book
should be addressed to:

    Reader Inquiry Branch
    Editorial Department
    TAB BOOKS Inc.
    P.O. Box 40
    Blue Ridge Summit, PA 17214

Cover photograph courtesy of E and L Instruments/Global Specialties (Interplex Electronics Companies).

Contents

Preface to the Second Edition

The first edition of this book was immensely popular because it addressed a need of electronic servicers to learn digital electronics—at least six printings proved that the need was present. Several years later, the need still exists, and to answer that need this second edition was produced.

Much of the text in the second edition was in the first edition, but several oversights were corrected. Most notable is the coverage of digital and other forms of clock circuits. These clocks are oscillators or multivibrators, which were originally covered only under the rubric of timers. The chapter on clocks in this edition is large, and it covers many forms of clock circuits. Although some of them do not appear to have immediate applications in digital circuits, they are included to teach the basic principles governing the whole class of oscillators.

Also expanded is the coverage of microprocessors. When the first edition was written, the microprocessor was relatively new to most readers, and only a few advanced "hackers" were aware of those machines. Today, we have microprocessor controlled cassette players, VCRs, television sets, and automobile ignition/fuel systems. Clearly, the time of the microprocessor has arrived, so this expansion was needed.

I hope you get as much out of this book as other readers got out of the first edition.

Introduction

Digital electronics was once a very exclusive and limited field. Only a handful of technicians engaged in industrial, military, or computer electronics had to know anything at all about digital circuits. Even those few found that much of their daily work was, in reality, *analog* electronics rather than digital. Many industrial and military control systems, for example, were nothing more than dedicated, or preprogrammed, analog computers that performed a special function. But today, digital is everywhere. Computers, for example, are now cheaper, faster, and easier to operate than ever before. The early machines were dinosaurs compared with today's versions. My personal microcomputer, which easily fits on a desk top along with the typewriter used to type this book, has more computing power and is faster than the $156,000 computer of 10 years ago, yet is *150 times cheaper.*

Point of sale terminals are sprouting up all over the place. They will verify credit and checks and allow electronic transfer of funds from a customer's account to a merchant's. Bounced checks due to insufficient funds are becoming extinct.

And look at the consumer market. Microprocessor control of microwave ovens, home heating and air conditioning, TV and FM channel selectors, at least one car radio tuner, a large pile of electronic games, and a host of other devices are digital.

In short, anyone in electronics at all is or soon will be seeing digital electronic circuits. Even technicians who have been in the trade for several decades must learn "digital." Younger technicians are finding more and more emphasis placed on digital electronics in their technical schools.

This is a self-study guide primarily aimed at technicians already in the field of electronics who want to learn something about digital circuits, possibly to save their job next year. It is also suitable for hobbyists, amateur radio operators, and technical school students who have already studied electronic basics.

Chapter 1

Overview of Digital Electronics

It is difficult to say where digital electronics began, but it is easy to say what it has become. Some engineers are fond of proclaiming that digital is the way of the future, and they may well be right. Note well, however, that it is a big mistake to count analog electronics as "down and out," because many electronics jobs are still best done by analog circuits. Many, perhaps most, transducers used in industrial applications, for example, are analog in nature.

Despite the fact that analog is alive and well, digital has become a way of life in many areas of the electronics industry. People who once did not need to consider digital electronics at all, are now finding themselves immersed in it quite deeply. Many consumer electronic products, for example, now use digital circuits. Several television receivers use digitally-controlled phase-locked loop (PLL) tuners for VHF/UHF channel selection and other digital circuits to display data such as the channel number, date, or time of day. Many high-fidelity FM tuners on the market use a microprocessor and many digital car radios also use PLL techniques.

Tremendous growth has occurred in those areas of electronics that have traditionally seen a heavy digital involvement. Accelerated by the introduction of the microprocessor, this growth has caused some pundits and prognosticators to call it a new industrial revolution.

Look at the computer market as an example. From a handful of machines in the era immediately following World War II, the

1

computer industry has grown to the point where every medium-sized business can afford a machine. Microcomputers can now do jobs for $2000 that only a decade ago required almost $100,000 worth of equipment! I find myself in awe, for example, of the fact that my own first desk top microcomputer was put together for less than $1000 and is more powerful than the $50,000 machine used in engineering school during the late '60s.

The growth of microcomputers has helped to produce relatively sophisticated machines in a price range where doctors, dentists, lawyers, and other small business people can afford them. The system that now costs the same as a new car, once cost $150,000 and required an entire room to house it. Now, a simple desktop IBM-PC or MacIntosh is sufficient. There has also grown up an industry of computer-related digital industries and industries that were not electronic at all until "digital" came along at cheap prices.

Products related to computers include cathode ray tube (CRT) video terminals, which are familiar to television repair technicians, and special point-of-sale computer terminals for retail establishments. The latter only superficially seem to appear like computer equipment and more nearly resemble calculators. The point-of-sale terminal may be little more than a special keyboard and a data communications device called a *modem* (*mo*dulator/*dem*odulator), or if may contain a real microprocessor device. Examples of these machines can be seen at almost any department store and supermarket cashier. Some grocery stores use special digital bar code readers and inventory control computers, and at least some fast food chains use special point-of-sale terminals in which the keys are labeled with the names of the food items offered. If an employee presses "3" and "hamburger," the internal computer automatically totals up the price of three hamburgers, calculates and adds the tax, and then displays the price to the customer. These special terminals allow the merchant to use low-paid, unskilled labor in the retail operation, rather than more skilled people capable of operating the actual computer. What has worked for fast food chains also works in most other forms of mass merchandising. And it allows interrogation of credit account data and electronic transfer of funds, as well.

In the field of devices that were not previously digital—or even electronic—the electronic cash register has replaced the former mechanical monster. Also seen are produce/meat scales that are legal for trade and home bathroom scales that are not legal for trade but tell all too well if someone has been "illegal" on his diet. I recently bought gasoline in a self-service station that has a pump with an electronic digital readout—and an electronic ticket printer!

The upshot of all this activity is a terribly large increase in digital devices in the market place. Yet service capability has not kept pace with sales and introduction of new devices. The editor of a popular electronic servicing industry magazine (a technician's trade journal) once went to an electronics show and asked several displayers of electronic digital products who was going to service their wares. Their answers were all the same: television service technicians and others already in the business of electronic service. A number of these companies, especially the smaller ones, cannot afford to locate a service technician or field engineer within one hour's drive from every potential customer. Their only options are to require customers to send the product back to the factory or a regional service center, or appoint local servicers to do the job. Clearly, then, both market expansion and the inclusion of digital devices in previously nondigital or nonelectronic products creates a golden opportunity for people who want to get into digital servicing.

WHAT IS DIGITAL ELECTRONICS?

Digital electronic circuits differ from so-called analog circuits in that they are *binary* in nature; that is, they have only *two* possible states. These states are symbolically represented by the digits "0" and "1" of the binary (base 2) number system, and they are actually two different voltage levels.

In most, but certainly not all, the 0 and 1 states represent voltage levels of 0 and $+2.4$ to $+5$ volts, respectively. These levels have become standard because they were used in the immensely popular transistor-transistor logic (TTL) integrated circuit devices. Modern complementary metal oxide semiconductor CMOS ICs, however, may use 0 and $+5$ volts, 0 and any positive potential between $+4$ and $+18$ volts, or a negative voltage for logical 0, and a positive voltage for logical 1 (-6 volts for logical 0 and $+6$ volts for logical 1 is popular).

Before proceeding further, however, let's first dispense with a little matter of notation. In order to avoid confusion, let us use the term LOW to denote logical 0, and HIGH to denote logical 1. This convention will be used every place in this book, unless we actually intend to denote the binary digits 1 and 0.

Let us further define digital signals by comparing three types

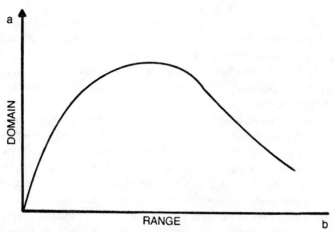

Fig. 1-1. Continuous analog signal.

of signals: *analog, sampled analog,* and *digital.* These are illustrated in Figs. 1-1 through 1-3.

The analog signal is shown in Fig. 1-1. Note that any signal can be defined as values within a *domain* (vertical axis) and a range (horizontal axis). In electrical circuits, these are *amplitude* and *time,* respectively. The analog signal is *continuous,* meaning that it can take on any amplitude from 0 to *a,* and it is defined at any instant in time over the interval 0 to *b.*

The analog signal, then, is defined for all values possible within the range and domain. This is not so with the sampled signal, however. It is defined for any amplitude from 0 to *a* but is only

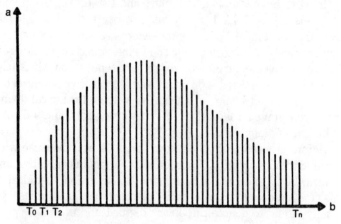

Fig. 1-2. Sample analog signal.

4

Fig. 1-3. Digitized signal.

defined for certain *specific* points in time. The value of the amplitudes in Fig. 1-2 are known only at times T_0, T_1, T_2 ... T_n, but they are not known at instants between these times. This is not to say, however, that we cannot *infer* probable values at other instants from the overall shape of the curve. The method of inferring these unknown values is valid only if the sampling rate (number of samples per second) is greater than twice the highest frequency component present in the waveform being sampled.

A *digital* signal is defined for only specific values of amplitude and at specific instants in time. Furthermore, the amplitude values will be given in one of several binary codes (see Chapter 3) instead of actual voltage values. Although a graph is shown in Fig. 1-3, the binary values could just as easily be given in tabular form. Both forms convey the same information in binary systems.

Note that all three types of signal can denote the same information, but only the digital signal can be used in computers and other digital circuits. The other two forms must be first converted to digital form before being input to a digital instrument or computer.

DIGITAL UNMASKED

To many people whose sophistication is in nondigital areas of electronics, digital circuits often present something of a mystery. This was probably started, and is certainly sustained, by the computer industry. In reality, though, most digital electronic circuits are less difficult to understand than the workings of a color television receiver! Also, fewer technologies exist in digital equipment than in color TV. Digital equipment tends to use only digital electronics, while modern color TVs might use digital, VHF/UHF rf, pulse circuits, audio circuits, and a good mix of transistors and other discrete devices, linear ICs, and digital ICs.

No one who is competent to service color TV receivers, two-way radios, or other electronic devices, need be afraid to service digital electronics. All of the concepts and devices are easy to understand and are imminently knowable.

Certain experiments in this book will aid you in learning the materials presented in the text. They are designed to give you a bit of "hands on" experience as you study the basics. Every effort has been made to select IC devices for these experiments that are widely available in hobby electronic stores, local electronic parts wholesalers, or mail-order electronic parts distributors. The latter may be found through their advertisements in publications such as *Radio-Electronics, QST, Byte, 73, Ham Radio*, etc. The experiments have also been designed around test equipment normally found in any properly equipped color television repair shop; no high-priced logic analyzers, or 200-MHz oscilloscopes, will be needed.

It is desirable that you buy a digital logic breadboard such as the Heath or the AP Products *Powerace 102*. The switch, pin, and LED indicator designations used in the experiments in this book are for the *Powerace 102*.

Chapter 2

Number Systems

You're going to learn a lot about various number systems in this chapter. Specifically, you'll learn about the binary number system, the notation of the octal and hexadecimal number systems, the principles of binary arithmetic, and the conversion of numbers in one system to equivalent numbers in another system.

SELF-EVALUATION QUESTIONS

These questions test your prior knowledge of the material in this chapter. Try answering them before you read the text. Then look for the answers as you read. When you are finished studying this chapter, try again to answer the questions. If you cannot answer a particular question, place a check mark beside it, and go back and reread appropriate portions of the text. When you are finished, try answering the questions at the end of the chapter using the same technique.

- ☐ Does $101_{10} = 101_2$?
- ☐ Convert $3F_{16}$ to octal notation.
- ☐ Binary, octal, decimal and hexadecimal are examples of number systems.
- ☐ What is the decimal equivalent of 362_8?
- ☐ Describe the procedure for binary subtraction using *twos complement* notation.

INTRODUCTION TO NUMBER SYSTEMS

One of the first complaints heard from new students of digital electronics is over the matter of learning the *binary* (base 2) number system. But this is highly desirable, and one step further, it is also essential that you become at least familiar with the *octal* (base 8) and *hexadecimal* (base 16) number systems. The two latter systems are frequently used in computer work as a shorthand for the binary numbers actually used in the computer. For example, binary 01001110_2 can also be equivalently represented as hexadecimal $4E_{16}$ or octal 116_8.

We all grew up using the *decimal*, or base 10, number system, so it is seemingly simpler to us. But it is, in reality, no easier than base 2, base 8, or base 16, or even base 99, for that matter. The same basic concepts used in decimal numbers are also found in binary, octal, and hexadecimal. Only the specific digits change between the number systems; the concepts and ideas remain the same. A comedian/mathematician from an early '60s television show used to claim that "base 8 is exactly like base 10 if you're missing two fingers!" While that statement was designed to raise laughs from parents of children struggling with "new math," it is nonetheless true. Similar statements could easily be made of the other bases. Of course, with hexadecimal, you would need *extra* fingers or use some of your toes.

Because several base number systems are used in this text, we will adopt a special notation. The base will be given as a subscript. Therefore, "101_2" means "101" in the binary (base 2) system, while "101_8" means "101" in the octal (base 8) number system, "101_{10}" is "101" in the decimal (base 10) system, and "101_{16}" is "101" in hexadecimal (base 16). Some systems us "\$" or "H" to denote hexadecimal (e.g., "$2B_{16}$" = "2BH" or "\$2B").

Decimal Numbers

Before tackling number systems in other bases, let us first review the basic concepts of all number systems, using the familiar base 10, or decimal, number system as an example. If you passed second grade, then you learned this system. And even if you did not get past that grade, but are able to figure out if the clerk who sold you this book gave the correct change, then you still learned the decimal number system. After relearning the basics, you will tackle the binary, octal, and hexadecimal number systems.

All four number systems that we have mentioned (binary, octal, decimal, and hexadecimal) are examples of *weighted* number systems. The actual value of a number is dependent upon the *digits* used, such as 0, 1, 2 . . . , and their respective *position* with respect to each other.

The decimal, or base 10, number system has 10 digits: 0, 1, 2, 3, 4, 5, 6, 7, 8, and 9. These digits each represent different quantities, or values that are familiar to you already.

With only 10 digits, however, we can represent only 10 different quantities. But instead of creating a different digit—a digit is merely a *symbol*—to represent each possible quantity from zero to infinity, *positional notation* is used. This means that a digit's position relative to other digits confers added value. For example, the decimal number 682_{10} means:

$$682_{10} = (6 \times 10^2) + (8 \times 10^1) + (2 \times 10^0)$$

or

$$682_{10} = (600) + (80) + (2)$$

The base of the number system (10, in this case) is called the *radix*. In general, all weighted number systems use a *digit* multiplied by a *radix raised to some power* at each position. The general form for all weighted number systems is:

$$D_nR^n + \ldots + D_3R^3 + D_2R^2 = D_1R^1 + D_0R^0$$

where the D terms are the digits (0 through 9 in base 10), and the R terms are the radix (10 in base 10).

When we place our example, 682_{10}, in the general form, it is

$$(0 \times 10^n) + \ldots + (0 \times 10^3) + (6 \times 10^2) + (8 \times 10^1) + (2 \times 10^0)$$

Note: Any number raised to the *zero* power, such as 10^0, is equal to 1. So $10^0 = 2^0 = 8^0 = 16^0 = N^0 = 1$.

We can use the general form equation to represent either whole or partial quantities. A *partial quantity* is a fraction. A decimal point must be used to separate the two. The digits to the left of the decimal point represent whole quantities, while those to the right represent fractional quantities. For example, 18.23 means:

$$18.23_{10} = (1 \times 10^1) + (8 \times 10^0) + (2 \times 10^{-1}) + (3 \times 10^{-2})$$

$$= \ (10) + (8) + (2 \times \frac{1}{10}) + (3 \times \frac{1}{100})$$

$$= \ 10 + 8 + \frac{2}{10} + \frac{3}{100}$$

Binary Numbers

The binary number system is just like the decimal number system if you're missing eight fingers. Such a situation would leave you two fingers, or two digits, which could be used for counting. These digits are 0 and 1. In the notation of the general form equation, then, a weighted binary number system would be

$$D_N 2^N + \ldots + D_3 2^3 + D_2 2^2 + D_1 2^1 + D_0 2^0$$

where the D terms are the digits (0 and 1 in binary).

For example, the binary number system number 01011_2 is the shorthand way of writing:

$$(0 \times 2^4) + (1 \times 2^3) + (0 \times 2^2) + (1 \times 2^1) + (1 \times 2^0) \ =$$
$$(0) + (8_{10}) + (0) + (2_{10}) + (1_{10}) \ = \ 11_{10}$$

At this point, we'll illustrate why the subscript notation must be used to let you know which system is being used. The digits "10" do not represent the same quantity in binary and decimal systems:

$$10_2 \ = (1 \times 2^1) + (0 \times 2^0) = 2_{10}$$
$$10_{10} = (1 \times 10^1) + (0 \times 10^0) = 10_{10}$$

Obviously, then, 10_2 does not equal 10_{10}.

The term *bit* used in digital electronics means *b*inary dig*it* and is the smallest unit of data possible. A bit can be either of the binary digits, 0 or 1. A *byte* is an array of *eight bits* arranged in a positional number representation from 00000000 to 11111111.

□ Exercise 1-1:

Find the *decimal* values of the following binary numbers. Use the notation of $D_N 2^N + \ldots + D_3 2^3 + D_2 2^3 + D_1 2^1 + D_0 2^0$.
For instance,

$$101_2 = (1 \times 2^2) + (0 \times 2^1) + (1 \times 2^0)$$
$$= 2^2 + 0 + 2^0 = 4 + 0 + 1 = 5_{10}$$

a. 101101   j. 0010
b. 110   k. 0011
c. 111111   l. 0100
d. 0010111   m. 0101
f. 100110   n. 0110
g. 1001111   o. 0111
h. 0000   p. 1000
i. 0001   q. 1001

Octal Numbers

The octal number system is just like the decimal number system. This time, though, you're missing two fingers. The octal number system uses the digits 0, 1, 2, 3, 4, 5, 6 and 7. Using the notation of the general form equation, a weighted octal number would look like:

$$D_N 8^N + \ldots + D_3 8^3 + D_2 8^2 + D_1 8^1 + D_0 8^0$$

where the D terms are any of the digits 0 through 7. In the notation of this equation, then, 362_8 would be:

$$(3 \times 8^2) + (6 \times 8^1) + (2 \times 8^0) =$$
$$(3 \times 64) + (6 \times 8) + (2 \times 1) =$$
$$(192) + (48) + (2) = 242_{10}$$

☐ **Exercise 2-2:**
Using the notation of this previous equation, find the decimal value of the octal numbers given below.

a. 312_8 i. 72_8
b. 377_8 j. 544355_8
c. 201_8 k. 011110_8
d. 110_8 l. 324_8
e. 35421_8 m. 327_8
f. 672_8 n. 666_8
g. 5012_8 o. 721_8
h. 51754_8 p. 345_8

Hexadecimal Numbers

The hexadecimal number system could be just as simple as the decimal number system if you could grow five extra fingers. The hexadecimal—nicknamed hex—system uses the 16 digits 0-15. But because all digits greater than 9 are double digits (10, 11, 12, 13, 14, and 15), and a single symbol is needed to represent each digit, the letters A through F to represent the digits 10 through 15. The digits of the hexadecimal number system are: 0, 1, 2, 3, 4, 5, 6, 7, 8, 9, A, B, C, D, E, and F, where A = 10, B = 11, C = 12, D = 13, E = 14, and F = 15.

Then a hexadecimal number would look like:

$$D_N 16^N + \ldots + D_3 16^3 + D_2 16^2 + D_1 16^1 + D_0 16^0$$

The hex number $7F6D_{16}$, then, is equal to

$$
\begin{aligned}
7F6D_{16} &= (7 \times 16^3) + (F \times 16^2) + (6 \times 16^1) + (D \times 16^0) \\
&= (7 \times 4096) + (15 \times 256) + (6 \times 16) + (13 \times 1) \\
&= 28{,}672 + 3840 + 96 + 13 = 32621_{10}
\end{aligned}
$$

□ **Exercise 2-3:**

Find the decimal value of the following hexadecimal numbers. Use the equation, $D_N 16^N + \ldots + D_3 16^3 + D_2 16^2 + D_1 16^1 + D_0 16^0$.

a.	FF_{16}	i.	$1B_{16}$
b.	$ABCDEF_{16}$	j.	$2C_{16}$
c.	$7F2_{16}$	k.	$4BB_{16}$
d.	$3A1_{16}$	l.	BBC_{16}
e.	$3A7B_{16}$	m.	377_{16}
f.	$4E_{16}$	n.	$A7F_{16}$
g.	$6F_{16}$	o.	$A6_{16}$
h.	$9C_{16}$	p.	$B2_{16}$

NUMBER SYSTEM CONVERSIONS

It sometimes becomes necessary to convert from one number system to another. This is often done by service technicians because they must correlate numbers given in the documentation for a piece of equipment. A manual for a computer, for example, might tell you to examine the eight-digit binary number stored in "memory location 372_8." But when you look at the front panel of the machine, it becomes immediately apparent that *binary* switches are the only means for calling up the specified address. A specialized pocket calculator made by Texas Instruments, called the *programmer,* will convert the numbers for you. Alternatively, for the numbers 0 through 33, use Table 2-1. But for all other numbers, or when tables and calculators are not available, use the following procedure.

This procedure converts all numbers to decimal form:

1. Write down the number being converted.
2. Multiply the most significant digit (the left-most) by the radix of the number.
3. Add the result of step No. 2 to the next most significant digit (MSD).
4. Multiply the result obtained in step No. 3 by the radix.
5. Repeat steps 3 and 4 until all digits are exhausted. The final result is the answer.

□ **Example 2-1:**
Convert 1101101_2 to decimal form.

1101101	Multiply MSD by radix	$1 \times 2 =$	2
1101101	Add result to next digit	$2 + 1 =$	3
	Multiply result by radix	$3 \times 2 =$	6
1101101	Add result to next digit	$6 + 0 =$	6
	Multiply result by radix	$6 \times 2 =$	12
110111	Add result to next digit	$12 + 1 =$	13
	Multiply result by radix	$13 \times 2 =$	26
1101101	Add result to next digit	$26 + 1 =$	27
	Multiply result by radix	$27 \times 2 =$	54
1101101	Add result to next digit	$54 + 0 =$	54
	Multiply result by radix	$54 \times 2 =$	108
1101101	Add result to next digit	$108 + 1 =$	109_{10}

□ **Example 2-2:**
Find the decimal value of 234_8.

234_8	Multiply the MSD by radix	$2 \times 8 =$	16
234_8	Add result to next digit	$16 + 3 =$	19
	Multiply result by radix	$19 \times 8 =$	152
234_8	Add result to next digit	$152 + 4 =$	156_{10}

Table 2-1. Digital Codes.

Decimal	Hexadecimal	Octal	Binary	Decimal	Hexadecimal	Octal	Binary
0	0	0	0	16	10	20	10000
1	1	1	1	17	11	21	10001
2	2	2	10	18	12	22	10010
3	3	3	11	19	13	23	10011
4	4	4	100	20	14	24	10100
5	5	5	101	21	15	25	10101
6	6	6	110	22	16	26	10110
7	7	7	111	23	17	27	10111
8	8	10	1000	24	18	30	11000
9	9	11	1001	25	19	31	11001
10	A	12	1010	26	1A	32	11010
11	B	13	1011	27	1B	33	11011
12	C	14	1100	28	1C	34	11100
13	D	15	1101	29	1D	35	11101
14	E	16	1110	30	1E	36	11110
15	F	17	1111	31	1F	37	11111
				32	20	40	100000
				33	21	41	100001

□ Example 2-3:

Find the decimal equivalent of $7F3A_{16}$.

$7F3A_{16}$	Multiply the MSD by the radix	$7 \times 16 = 112$
$7F3A_{16}$	Add result to next digit	$112 + 15 = 127$
	Multiply result by radix	$127 \times 16 = 2032$
$7F3A_{16}$	Add result to next digit	$2032 + 3 = 2035$
	Multiply result by radix	$2035 \times 16 = 32560$
$7F3A_{16}$	Add result to next digit	$32560 + 10 = 32570_{10}$

The preceding procedure converts any number in any base, R, to the equivalent number in decimal notation. The following procedure does just the opposite and converts a decimal number to an equivalent in base R.

1. Divide the decimal number by the radix of the new number system.
2. The remainder from this step becomes the least significant digit (LSD) of the equivalent number in base R.
3. Divide the quotient from step No. 1 by the radix.
4. The remainder from this operation becomes the next digit of the new number system.
5. Continue this process until the digits are exhausted.

□ Example 2-4:

Find the binary equivalent of 109_{10}.

Operation	Remainder
2/ 109	—
2/ 54	1 LSD
2/ 27	0
2/ 13	1
2/ 6	1
2/ 3	0
2/ 1	1
0	1MSD

This result agrees with that of Example 2-1

$$109_{10} = 1101101_2$$

□ **Example 2-5:**
Convert 156_{10} to octal.

Operation	Remainder
8/ 156	4 LSD
8/ 19	3
8/ 2	2 MSD
0	

This result agrees with Example 2-2.

$$156_{10} = 234_8$$

□ **Example 2-6:**
Find the hexadecimal equivalent to $32,570_{10}$.

Operation	Remainder	
	Dec.	Hex.
16/ 32570	10	A LSD
16/ 2035	3	3
16/ 127	15	F
16/ 7	7	7 MSD
0		

This result agrees with Example 2-3.

$$32570_{10} = 7F3A_{16}$$

These procedures allow you to convert into decimal form from radix R, or from decimal into radix R. It is usually less trouble to use both procedures when converting from radix R1 to radix R2, when neither R1 nor R2 are decimal. For example, when converting from hex to octal, it may be better to convert from hex to decimal, and then decimal to octal.

□ **Example 2-7:**
Find the hex equivalent of 377_8.

1. *First* find the decimal equivalent of 377_8.

    377_8 Multiply MSD by radix $3 \times 8 = 24$

377_8	Add result to next digit	$24 + 7 = 31$
	Multiply result by radix	$31 \times 8 = 248$
377_8	Add next result to next digit	$248 + 7 = 255$

The decimal equivalent of octal 377_8 is 255_{10}.

2. Now, find the hexadecimal equivalent to 255_{10}.

$$
\begin{array}{rl}
16/\underline{\;255\;} & \\
16/\underline{\;15\;} & 15 \quad \text{LSD} \\
0 & 15 \quad \text{MSD}
\end{array}
$$

The hexadecimal notation for the digit "15_{10}" is F, so the answer is FF_{16}. We now know that the hexadecimal equivalent of 377_8 is FF_{16}:

$$377_8 = 255_{10} = FF_{16}$$

BINARY ARITHMETIC

Binary arithmetic is used in computers and is also very handy to use when trying to figure out almost any complex digital instrument. With more and more instruments on the market employing the microprocessor chip, the understanding of binary arithmetic techniques is even more critical.

Addition

The rules governing binary *addition* are simple; some people believe that they are simpler than decimal arithmetic! The rules are as follows:

$$
\begin{aligned}
0 + 0 &= 0 \\
1 + 0 &= 1 \\
0 + 1 &= 1 \\
1 + 1 &= 0 \text{ plus carry 1}
\end{aligned}
$$

□ **Example 2-8:**
Add the binary numbers 01001_2 and 01110_2.

$$
\begin{array}{r}
1 \\
0\,1\,0\,0\,1 \\
+\ 0\,1\,1\,1\,0 \\
\hline
1\,0\,1\,1\,0
\end{array}
$$

□ **Example 2-9:**
Add the binary numbers 00001_2 and 01001_2.

$$
\begin{array}{r}
1 \\
0\ 1\ 0\ 0\ 1 \\
+\ 0\ 0\ 0\ 0\ 1 \\
\hline
0\ 1\ 0\ 1\ 0
\end{array}
$$

Subtraction

The rules for straight binary *subtraction* are as follows:

$$
\begin{aligned}
0 - 0 &= 0 \\
0 - 1 &= 1 \text{ borrow } 1 \\
1 - 0 &= 0 \\
1 - 1 &= 0
\end{aligned}
$$

Note, however, that it is difficult to build a digital electronic circuit that performs straight binary subtraction operations. It is, however, relatively trivial to build binary *adder* circuits. In most digital devices that are required to subtract, therefore, it is simpler to use *twos' complement* arithmetic to fool the circuit into thinking that it is adding!

Before discussing subtraction by twos' complement, let's find out what "twos' complement" is. We begin by telling you about the ones' complement of a binary number. The complement of any binary number is its inverse; i.e., the complement of 1 is 0, and the complement of 0 is 1. To form the ones' complement of any binary number, therefore, we change all of the 1s to 0s and change all of the 0s to 1s.

Digit	Complement
0	1
1	0

For example, the complement of 10110011 is 01001100. Note that all ones became zeroes and all zeroes became ones.

□ **Example 2-10:**
Find the complement of 1011001_2.

number	1011001
complement	0100110

The twos' complement of any binary number is found by adding 1 to the ones' complement of that number.

☐ **Example 2-11:**

Find the twos' complement of 1011001_2.

Number	1 0 1 1 0 0 1
Ones' complement	0 1 0 0 1 1 0
Add 1	+1
	0 1 0 0 1 1 1

0100111_2 is the twos' complement of 1011001_2.

Twos' Complement Subtraction

You can subtract a binary subtrahend from a binary minuend by *adding* the twos' complement of the subtrahend to the minuend. Keep in mind that subtraction can be viewed as adding a negative number, and that twos' complement notation is a form of representation for negative binary numbers. The following are several examples of twos' complement arithmetic, given several different types of situations that could arise.

Before studying the examples, however, a standard convention concerning the use of "+" and "−" signs is important. The last digit, separated from the number by a comma (,), will be used to denote the *sign* of the number. A 0 will represent positive (+) numbers, while a 1 will represent negative (−) numbers. See Fig. 2-1.

☐ **Example 2-12:**

Perform subtraction when minuend and subtrahend are *both positive*, and the minuend is *larger* than the subtrahend.

Subtract $0,100_2$ (4_{10}) from $0,111_2$ (7_{10}).

Minuend	0,111	+7
Subtrahend	− 0,100	−(+4)
	?	?

1. Complement the subtrahend	0,100	
	1,011	
2. Add 1	+ 0,001	
	1,100	(−4)

3. Add twos' complement
   of subtrahend to the   0,111   +7
   minuend   + 1,100   +(−4)
                          1 ◄——————— 0,011   3

(carry 1 discarded)

The problem is worked correctly because $0,0011_2 = +3_{10}$.
(Notice that the carry one is generated, but in this case it is
discarded or ignored).

□ **Example 2-13:**
Perform subtraction of a *negative subtrahend* from a *positive
minuend* when the minuend is *larger* than the subtrahend.

Subtract $1,001_2$ (i.e. $− 1_{10}$) from $0,011_2$ $(+3)$.

$$\begin{array}{cc} 0,111 & +3 \\ - 1,001 & - (-1) \\ \hline ? & ? \end{array}$$

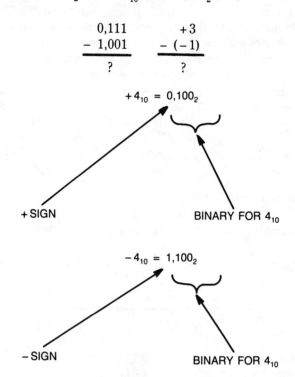

Fig. 2-1. The zero represents positive numbers, while the one represents
negative numbers.

20

Note in the decimal version that we are subtracting a negative quantity, so the problem is the same as adding their absolute values:

$$
\begin{array}{rr}
0,011 & +3 \\
+\ 0,001 & +1 \\
\hline
0,100 & +4
\end{array}
$$

We know the problem is correctly worked because $0,100_2 = +4_{10}$.

Multiplication

Binary multiplication can be done using the same method as decimal multiplication, or by *repetitive addition*. The usual method by pencil and paper is shown in Example 2-13.

☐ **Example 2-14:**

Multiply 1101_2 (D_{16} or 13_{10}) by 0100_2 (4_{16} and 4_{10}).

$$
\begin{array}{c}
1101_2 \\
\times\ 0100_2 \\
\hline
0000_2 \\
0000_2 \\
1101_2 \\
0000_2 \\
\hline
0110100_2
\end{array}
\qquad
\begin{array}{c}
D_{16} \\
\times\ 4_{16} \\
\hline
34_{16}
\end{array}
\qquad
\begin{array}{c}
13_{10} \\
\times\ 4_{10} \\
\hline
52_{10}
\end{array}
$$

Multiplication is nothing more than addition performed over and over again. For example, 4×5 is

$$4_{10} \times 5_{10} = 4 + 4 + 4 + 4 + 4 = 20_{10}$$
$$\text{5 repetitions of } +4$$

The same thing can be done with binary numbers; the technique is called *repetitive addition* or the *shift left* method. The technique is:

1. Set the answer initially to zero; i.e., 00000000_2.
2. Write down the multiplication so that the least significant

digit of the accumulated answer is aligned with the LSD of multiplicand.

3. Inspect the multiplier beginning with the LSD. If it is a 1, then add the multiplicand to the answer. But if it is a 0, then add 0 to the answer. Shift the multiplicand one bit to the left.

4. Repeat steps 2 and 3 until finished.

Examine the process by performing the same problem as in Example 2-15 to see if we get the same answer.

□ **Example 2-15:**
Use repetitive addition to solve Example 2-14.

$$\begin{array}{r} 1101_2 \\ \times\, 0100_2 \\ \hline ? \end{array}$$ Multiplicand
Multiplier

1. Tentative answer
   Multiplicand

   $$\begin{array}{r} 00000000000 \\ 1101 \\ \hline \end{array}$$

2. Test LSD of multiplier for 1/0. It is 0, so answer remains unchanged.

   010**0**

   ↙ LSD is 0

3. Tentative answer
   Multiplicand *shifted* 1 bit to the left

   $$\begin{array}{r} 00000000000 \\ 1101 \\ \hline \end{array}$$

4. Test (LSD + 1) of multiplier for 1/0. It is 0, so answer is unchanged.

   01**0**0

   ↙ LSD + 1 is 0

5. Tentative answer
   Multiplicand again shifted 1 bit to the left.

   $$\begin{array}{r} 00000000000 \\ 1101 \\ \hline \end{array}$$

6. Test (LSD + 2) of multiplier for 1/0. It is 1, so *add* multiplicand to tentative answer.

   0**1**00

   ↙ LSD + 2 is 1

22

7. New tentative answer   00000110100
   Multiplicand shifted   1101
   1 bit to the left.

8. Test multiplier
   for 1/0. It is   0100
   0, so answer
   remains the same   LSD + 3 is 0

9. There are no further digits in this binary number, so the
   answer is the same as in step 7: 00000110100.

   Because the zeroes to the left of the most significant 1 are
merely leading zeroes, drop them to shorten the answer to 110100.
   The repetitive addition method seems a little cumbersome. On
pencil and paper this is true, but in digital electronic circuits, adding
and left-shifting are very easy to accomplish. In fact, it is difficult
to do other operations, where addition and shifting are easy.

Logical Operations

   Digital circuits perform *logic operations* called NOT, OR, AND,
and Exclusive-OR (XOR). The OR and AND functions also have
inverted versions called NOR and NAND. In each case, two (or
more) bits are compared, and the logical outcome is determined
by which of these basic functions is being used.
   Again consider some matters of notation. The NOT function
is an inverting function. In the NOT function, a 1 will become 0,
and a 0 will become 1. We represent the NOT function as a symbol
with a bar overhead; i.e., $\overline{A}$ means the same as "NOT A." This
means that if A = 1, then the $\overline{A}$ value is 0. If A = 0, then $\overline{A}$ =
1. The NOT symbol is also used in conjunction with other functions,
as is the idea represented. A NOT gate connected with an OR gate
produces a NOR gate. Similarly, a NOT gate connected with an
AND gate produces a NAND gate.
   Multiplication (×) and addition (+) symbols are used to represent the AND and OR functions, respectively. A circled addition
symbol is used to represent the exclusive-OR (XOR) function. The
NAND and NOR functions are represented by the equivalent AND
and OR expressions with a NOT bar overhead. See Table 2-2.

RECAPITULATION

   Now go back to the beginning of the chapter and try answering
the self-evaluation questions. When you are finished, try answering

23

Table 2-2. Various Logical Operations.

NOT	NOT-A	$\overline{A}$	
AND	A AND B	$A \times B$*	
OR	A OR B	$A + B$	
XOR	A XOR B	$\underline{A + B}$	
NOR	A NOR B	$\overline{A + B}$	(A NOT-OR B)
NAND	A NAND B	$\overline{A \times B}$	(A NOT-AND B)

* Alternate form: $A \cdot B$

NOT Function.   Def.: Output is the inverse of the input.

$$
\begin{array}{cc}
A & \overline{A} \\
0 & 1 \\
1 & 0
\end{array}
$$

NAND Function. $A \times B$
Def.: Output is 1 only if both inputs are 1.

$$
\begin{aligned}
1 \times 1 &= 1 \\
1 \times 0 &= 0 \\
0 \times 1 &= 0 \\
0 \times 0 &= 0
\end{aligned}
$$

OR Function. $A + B$
Def.: Output is 1 if either input is 1

$$
\begin{aligned}
1 \times 1 &= 1 \\
1 \times 0 &= 1 \\
0 \times 1 &= 1 \\
0 \times 0 &= 0
\end{aligned}
$$

Exclusive-OR (XOR) $A + B$

Def.: Output is 1 if either input is 1, but not if both
   inputs are 1.

$$
\begin{aligned}
1 + 1 &= 0 \\
0 + 1 &= 1 \\
1 + 0 &= 1 \\
0 + 0 &= 0
\end{aligned}
$$

NOR Function. $\overline{A + B}$

Def.: Output is 0 if either input is 1.

$$
\begin{aligned}
\overline{1 + 1} &= 0 \\
\overline{0 + 1} &= 0 \\
\overline{1 + 0} &= 0 \\
\overline{0 + 0} &= 1
\end{aligned}
$$

the following questions and solving the following problems.

QUESTIONS

- [ ] Define the term *bit*.
- [ ] Define the term *byte*.
- [ ] The octal number system is base—_____.
- [ ] The decimal number system is base—_____.
- [ ] The hexadecimal number system is base—_____.
- [ ] The binary number system is base—_____.
- [ ] List all digits used in the binary number system.
- [ ] List all digits used in the decimal number system.
- [ ] List all the digits used in the octal number system.
- [ ] List all digits used in the hexadecimal number system.
- [ ] The *radix* of the binary number system is_____.
- [ ] List the rules for binary addition.
- [ ] List the rules for binary subtraction.

PROBLEMS

- [ ] Convert 101_2 to octal notation.
- [ ] Convert $3F_{16}$ to octal notation.
- [ ] Convert 110_2 to decimal.
- [ ] Convert 029_{16} to decimal.
- [ ] Convert 123_8 to decimal.
- [ ] Perform the addition below

$$11011_2$$
$$+ 101_2$$

- [ ] Use twos' complement notation to perform the following binary subtraction.

$$1101_2$$
$$- 101_2$$

Chapter 3

Digital Codes

_____

In this chapter, you'll learn the most common code schemes used in digital circuits to represent binary data. Also, you'll learn the types of devices sometimes used to generate or store codes.

SELF-EVALUATION QUESTIONS

These questions test your prior knowledge of the material in this chapter. Try answering them before you read the text and then look for the answers as you read. When you have finished studying this chapter, try again to answer these questions. If you cannot answer a particular question, place a check mark beside it and go back and reread appropriate portions of the text. When you are finished, try answering the questions at the end of the chapter using the same technique.

☐ The decimal equivalent of BCD 0100 0110 1001 is _____.

☐ Represent hexadecimal number FF A4 in split-octal notation.

☐ The _____ code changes only one bit for each increment.

☐ Three keyboard codes found in computer terminals are _____, _____, and _____.

A digital code is a binary representation of data. Nothing is mysterious about codes; they are merely methods for presenting data in formats that digital machines can easily use. A code may represent numerical, alphabetic, alphanumeric, control signals, etc. Most computer or teletypewriter keyboard codes, for example, contain numbers, alphabetic characters (upper and lower case), and control codes, such as carriage return.

One property of any useful machine code is that it is binary in nature, so that *on-off* switching circuits (discussed elsewhere in this book) can be used in the machine. It turns out that binary circuits are easiest to use in such cases.

Binary numbers are a form of "code" used to represent numerical quantities, but ordinarily, a binary code *does not* correspond to the binary number with the same bit structure. For example, in the ASCII code to be discussed shortly, the numerical character "7" is 00110111, while the binary number "7" (when expressed with the same number of bits) is 00000111. It is important to remember the difference between a *symbol* and a *number*. A computer may perform an arithmetic operation in which the result is the number seven (representing a quantity), but the output display device, which could be a printer or CRT video terminal, may require ASCII coding. If the programmer does not first convert the binary number 00000111 to 00110111, an incorrect result will be displayed to the outside world. In this particular example, it just happens that 00000111 in ASCII is the "BEL" symbol, so the signal bell of the printer would ring instead of character "7" being printed. It would be a rare situation, indeed, if the binary representation for a quantity was the same as the code for the character(s) that represents that quantity.

HEXADECIMAL CODE

A binary code of, say, four bits could be represented by a hexadecimal (base 16) code. This would greatly simplify entering data or instructions into a digital machine. Entering 7F, for example, is a lot easier than entering all eight binary digits, or 01111111, needed in the binary representation. Using hexadecimal notation permits four bits to be entered at a time. Let us say that we have an eight-bit binary number:

$$1\ 0\ 1\ 1\ 1\ 1\ 0\ 1$$

To enter this number on a bit-by-tedious-bit basis would mean using eight separate keystrokes: 1-0-1-1-1-1-0-1. But we can break the number 10111101 into two four-bit groups:

$$1\ 0\ 1\ 1 \qquad 1\ 1\ 0\ 1$$

Because 1011 is hexadecimal number B_{16} and 1101 is hex D_{16}, the hexadecimal number BD_{16} represents the binary pattern 10111101.

The hexadecimal system can also represent binary words that have bit lengths not divisible by four, provided that we assume the leading digits to be zeroes. For example, assume a ten-bit number:

$$1\ 0\ 0\ 0\ 1\ 0\ 1\ 0\ 1\ 1$$

We can represent this with three hexadecimal digits:

$$0\ 0\ 1\ 0 \qquad 0\ 0\ 1\ 1 \qquad 1\ 0\ 1\ 1$$

Fill in the extra positions with zeroes. Because 0010 is hexadecimal 2_{16} and 1011 is hexadecimal B_{16}, hexadecimal $22B_{16}$ represents the same quantity as binary 1000101011 (no extra zeroes this time).

SPLIT-OCTAL

We can also represent binary data in the *split-octal* system by grouping the binary bits in groups of not more than three each. for example, the binary code 10111101_2 was represented by hexadecimal BD_{16} in the previous section. This same binary number can be divided up into octal groups:

$$010 \qquad 1\ 1\ 1 \qquad 1\ 0\ 1$$

Fill in any leading zeroes that are needed. Because binary 010_2 is octal 2_8, binary 111 is octal 7_8, and binary 101_2 is octal 5_8, octal 275 produces the same binary bit pattern as binary number 10111101. In this case, however, only three keystrokes (2-7-5) are needed instead of eight (1-0-1-1-1-1-0-1). Both octal and hexadecimal notation can represent the numerical, alphabetic, and control codes of code systems that are commonly used on keyboards, printers, teletypewriters, etc.

BINARY-CODED DECIMAL

The *binary-coded decimal* (BCD) is used to represent the 10 digits

Table 3-1. Binary Coded Decimal.

Decimal	BCD	Decimal	BCD
0	0000	5	0101
1	0001	6	0110
2	0010	7	0111
3	0011	8	1000
4	0100	9	1001

of the decimal number system in a four-bit format. See Table 3-1.

BCD is merely another form of decimal notation that is compatible with computers and other digital circuits. BCD words are therefore grouped exactly the same as regular decimal digits. The actual value of a BCD word is determined by its position relative to the other words, in a power of 10 system. For example, see Table 3-2.

The BCD number system is difficult to use in making calculations, but it is not impossible. Its main use is in systems requiring a numerical output display, such as a frequency counter, digital panel meter, clock, scientific instrument, a toy, etc.

Several BCD decoder chips will convert a BCD input to an output capable of driving the principal display devices, which could be Nixie® tubes, seven-segment readouts, and so on. These are discussed in a later chapter.

EXCESS-3 CODE

The excess-3 code is formed by adding three to the BCD numbers. ($3_{10} = 011_2$). For example, see Table 3-3. Excess-3 code is used to make digital subtraction in BCD a little easier. Recall that complement arithmetic is often used with binary digits. This cannot be done in BCD, however, because it would occasionally result in a disallowed bit pattern that is not recognizable as any of the 10 BCD numbers.

Table 3-2. Examples of BCD.

Decimal	BCD
2	0010
23	00100011
175	000101110101

Table 3-3. The Excess-3 Code.

Decimal	BCD	Excess-3
0	0000	0011
1	0001	0100
2	0010	0101
3	0011	0110
4	0100	0111
5	0101	1000
6	0110	1001
7	0111	1010
8	1000	1011
9	1001	1100

GRAY CODE

Shaft encoders and certain applications work best if the binary code generated *changes only one bit at a time* for each change of state. An example of such a code is the Gray code, shown in Table 3-4. The Gray code is used most often where a transducer on a mechanical device is directly generating digital data, such as shaft position.

ALPHANUMERIC CODES

Actually, several different codes each represent alphabetic or numeric *characters*. Remember the distinction between characters and numbers—both represent quantities, but in different ways. We will discuss the obsolete but still found Baudot code, the popular ASCII code, and the IBM EBCDIC code.

Alphanumeric codes are used in cases where there is an output

Table 3-4. The Gray Code.

Decimal	Binary	Gray Code
0	0000	0000
1	0001	0001
2	0010	0011
3	0011	0010
4	0100	0110
5	0101	0111
6	0110	0101
7	0111	0100
8	1000	1100
9	1001	1101
10	1010	1111

display device, such as a CRT video terminal, teletypewriter, or computer printer. Standardization, at least to a few different codes, allows connection of devices by different manufacturers.

Baudot Code

One of the earliest machine codes in general use was the Baudot code (see Table 3-5). Used in most of the earlier models of teletypewriter machines, Baudot has been largely replaced by another, more modern code (ASCII). Many Baudot machines are still in daily use, but their numbers are dwindling. Note that most teletypewriters will have a keyboard with a shift key just like a regular typewriter. In most cases, the capital letters are on the shifted keyboard and the lower case letters are on the unshifted keyboard. But some teletypewriters are just the opposite.

The Baudot code uses five bits, so it's capable of representing only 2^5, or 32, different characters or control signals. A shift control, much like that found on a regular typewriter, allows another 32 characters, for a total of 64. Note that not all 64 characters are used on all Baudot-encoded teletypewriters.

ASCII Code

The American Standard Code for Information Interchange (ASCII) is the code most often used on computer and video terminal keyboards. It is especially common on computers. The ASCII code is a seven-bit code and can represent up to 2^7, or 128, different characters or control signals. Recall that this is twice the capacity of the five-bit Baudot machines.

An eight-bit format is often used in ASCII machines, especially those used with eight-bit microcomputers. The eighth bit is used as a parity indicator in some cases and as a strobe bit in others. A strobe bit, incidentally, tells the computer or other digital machine that the data on the seven data lines is valid and stable. Most machines are designed to ignore activity on the ASCII input line until a strobe bit is present.

In the ASCII listings shown in Table 3-6, the hexadecimal form of notation is used. The bit pattern for the specific character can be found by converting from hex to binary. An "&" sign is given in hexadecimal as 26_{16} and would be 0010 0110, or simply, 00100110_2.

EBCDIC Code

The Extended Binary Coded Decimal Interchange Code

Table 3-5. The Baudot Code.

B5	B4	B3	B2	B1	Regular	Shifted
0	0	0	0	0	Blank	Blank
0	0	0	0	1	E	3
0	0	0	1	0	linefeed	linefeed
0	0	0	1	1	A	-
0	0	1	0	0	space	space
0	0	1	0	1	S	Bell
0	0	1	1	0	I	8
0	0	1	1	1	U	7
0	1	0	0	0	Car. Ret.	Car. Ret.
0	1	0	0	1	D	$
0	1	0	1	0	R	4
0	1	0	1	1	J	'
0	1	1	0	0	N	.
0	1	1	0	1	F	!
0	1	1	1	0	C	:
0	1	1	1	1	K	(
1	0	0	0	0	T	5
1	0	0	0	1	Z	"
1	0	0	1	0	L	)
1	0	0	1	1	W	2
1	0	1	0	0	H	#
1	0	1	0	1	Y	6
1	0	1	1	0	P	0
1	0	1	1	1	Q	1
1	1	0	0	0	O	9
1	1	0	0	1	B	?
1	1	0	1	0	G	&
1	1	0	1	1	(figures)	(figures)
1	1	1	0	0	M	.
1	1	1	0	1	X	/
1	1	1	1	0	V	;
1	1	1	1	1	(letters)	(letters)

Fig. 3-1. Hollerith card.

33

Table 3-6. ASCII Code.

Hex Code	Meaning	Comments
00	NUL	null
01	SOH	start of heading
02	STX	start text
03	ETX	end text
04	EOT	end of transmission
05	ENQ	enquiry
06	ACK	acknowledgement
07	BEL	bell
08	BS	back space
09	HT	horizontal tab
0A	LF	line feed
0B	VT	vertical tab
0C	FF	form feed
0D	CR	carriage return
0E	SO	shift out
0F	SI	shift in
10	DLE	data link escape
11	DC1	direct control 1
12	DC2	direct control 2
13	DC3	direct control 3
14	DC4	direct control 4
15	NAK	negative acknowledgement
16	SYN	synchronous idle
17	ETB	end of transmission block
18	CAN	cancel
19	EM	end of medium
1A	SUB	substitute
1B	ESC	escape
1C	FS	form separator
1D	GS	group separator
1E	RS	record separator
1F	US	unit separator
20	(special)	-
21	!	-
22	"	-
23	#	-
24	$	-
25	%	-
26	&	-
27	'	-
28	(	-
29	)	-
2A	*	-
2B	+	-
2C	,	-
2D	-	-
2E	.	-
2F	/	-
30	0	-
31	1	-
32	2	-
33	3	-

Hex Code	Meaning	Comments
34	4	-
35	5	-
36	6	-
37	7	-
38	8	-
39	9	-
3A	:	-
3B	;	-
3C	>	-
3D	=	-
3E	<	-
3F	?	-
40	@	-
41	A	-
42	B	-
43	C	-
44	D	-
45	E	-
46	F	-
47	G	-
48	H	-
49	I	-
4A	J	-
4B	K	-
4C	L	-
4D	M	-
4E	N	-
4F	O	-
50	P	-
51	Q	-
52	R	-
53	S	-
54	T	-
55	U	-
56	V	-
57	W	-
58	X	-
59	Y	-
5A	Z	-
5B	[	-
5C	/	-
5D	]	-
5E	.	-
5F	|	-
60	^	-
61	a	-
62	b	-
63	c	-
64	d	-
65	e	-
66	f	-
67	g	-
68	h	-
69	i	-

Hex Code	Meaning	Comments
6A	j	-
6B	k	-
6C	l	-
6D	m	-
6E	n	-
6F	o	-
70	p	-
71	q	-
72	r	-
73	s	-
74	t	-
75	u	-
76	v	-
77	w	-
78	x	-
79	y	-
7A	z	-
7B	{	-
7C	|	-
7D	}	-
7E	~	-
7F	DEL	-

(EBCDIC) is most often associated with IBM Corporation equipment. This code can be most easily explained by referring to the Hollerith card shown in Fig. 3-1. This card was originally used to automate the tabulation of the 1890 United States census, yet remains popular today. It is, in fact, one of the mainstays of the data processing industry.

The Hollerith card is divided into rows and columns. There are 12 horizontal rows and 80 vertical columns. There are also two areas on this card. The rows numbered 0 through 9 form one area, while rows 11 and 12 form the other. Punches made in rows 11 and 12 are called *zone punches*.

Characters are represented by one-byte binary words. For this discussion, we can divide the byte into two *nybbles* of four bits each.

Table 3-7. Hexadecimal Notation.

Digit	Hex Code	Digit	Hex Code
0	FØ	5	F5
1	F1	6	F6
2	F2	7	F7
3	F3	8	F8
4	F4	9	F9

Table 3-8. EBCDIC Representation of the Alphabet.

Letter No.	1st Third	2nd Third	3rd Third
1	A	J	-
2	B	K	S
3	C	L	T
4	D	M	U
5	E	N	V
6	F	O	W
7	G	P	X
8	H	Q	Y
9	I	R	Z

This convention allows us to use hexadecimal notation. The code for digits is shown in Table 3-7.

Conversion to binary merely requires the substitution of the binary equivalents of the hexadecimal numbers. For example:

Decimal	EBCDIC	Binary
6	F6	11110110

EBCDIC representation of numbers places a hexadecimal F_{16} in the most significant position, so the four most significant binary digits will be 1111_2. Those positions are used when representing alphabetic characters and indicate which third of the alphabet the letter falls in. There are 26 letters in our alphabet, so the alphabet can be divided roughly into thirds. See Table 3-8.

Table 3-9. Alphanumeric Portion of the EBCDIC Code.

Character	Hex Code	Character	Hex Code
A	C1	O	D6
B	C2	P	D7
C	C3	Q	D8
D	C4	R	D9
E	C5	S	E2
F	C6	T	E3
G	C7	U	E4
H	C8	V	E5
I	C9	W	E6
J	D1	X	E7
L	D2	Y	E8
M	D3	Z	E9
N	D4		

Two punches are required to represent a letter on a Hollerith card. A zone punch tells us in which third of the alphabet the letter appears, and the numeric punch tells us the letter number. Letters in the first third of the alphabet have a row-12 zone punch, those in the second third of the alphabet have a row-11 zone punch, while those in the third third of the alphabet have a row-0 zone punch.

Consider, for example, the letter D. It is the fourth letter in the first third of the alphabet. It will be represented by a row-12 zone punch and a row-4 numeric punch, both in the same column. The letter N, on the other hand, is the fifth letter in the second third of the alphabet. It is, therefore, represented by a row-11 zone punch and a row-5 numeric punch.

A slight departure from the system is noted in the last third of the alphabet. There is no row-1 punch. The first letter in this group is S and it is represented by a row-0 zone punch and a row-2 numeric punch. The alphanumeric portion of the EBCDIC code is shown in Table 3-9.

Chapter 4

Digital IC Logic Families

Integrated circuits are used in all areas of electronics, especially digital electronics. In this chapter you will learn the different types of digital integrated circuits and how these ICs are used. Also, you'll learn the problems associated with each IC family and how they are avoided and the special handling procedures for certain ICs that are sensitive to static electricity damage.

WHAT ARE LOGIC FAMILIES?

A logic family is a group of IC devices manufactured using similar technology to possess similar attributes and work together using direct interconnection rather than elaborate interfacing circuitry.

Several different families and subfamilies have been developed over the past several decades: resistor-transistor logic (RTL), diode-transistor logic (DTL), transistor-transistor logic (TTL), emitter-coupled (AC) logic (ECL), high-threshold logic (HTL), complementary metal oxide semiconductor (CMOS) logic, and Schottky TTL. There are also several new subfamilies that evolved from early CMOS technology (e.g., NMOS, PMOS).

Some of these types of devices are considered obsolete, while others are still in current production. We will cover the obsolete devices briefly, however, because they may still be found in some older equipment, even though they are no longer incorporated into new designs.

Each of the currently popular families has its own place, and electronic circuit designers must make decisions based on what they perceive as the reasonable trade-offs between the vices and virtues of the respective families. One young fellow I know boasted to his very experienced electronics instructor that he could "do anything in CMOS that can be done in TTL, without all of that high current drain!" The instructor coolly asked, "Can you build an 80-MHz counter or shift register?" The answer is no, because CMOS is a slow-speed family.

Speed and *power consumption* tend to be directly proportional to each other in digital devices for good reason. High speed usually means increased power consumption. It seems that low power operation, with its reduced current drain, means higher internal resistances and usually higher internal capacitances. This makes the RC time constants of the circuit longer, thereby accounting for the slower speeds. When you reduce the impedances inside the device and the interelectrode capacitances in order to make the device faster, such as for operation at a higher frequency, then you also increase the current drain; hence, you increase the power consumption.

Before proceeding to a discussion of the various families of IC device, let's first dispense with a little housekeeping. What is meant exactly by the terms *current source, current sink,* and *fan-out/fan-in*?

A current source is a circuit that will supply current to the load. If you connect a resistor to one side of a power supply, then it becomes a current source to circuits connected between the free end of the resistor and the other side of the power supply.

A current sink, on the other hand, is a circuit that accepts current from another circuit. In the simplified example above, the load connected to the current source will act as a current sink.

Most IC logic families have inputs and outputs that are matched as to their current source or sink roles. A TTL input, for example, acts as a current *source*. The TTL output, on the other hand, acts as a current *sink*. When an output is used to drive an input, therefore, the current produced at the input will find a path to ground through the output of the driving device.

No matter how else you view them, IC digital devices are still little more than collections of transistors, diodes and, resistors. Just as in any direct-coupled transistor circuit, one must pay attention to the drive levels, impedance matching, and voltage/current levels at the interface between two stages. In most analog electronic circuits, the designer must do this each and every time a new circuit is designed, but the digital electronics industry has devised a method for specifying input drive requirements and output drive capabilities. The popular TTL family, for example, has standardized on a +5 VDC power supply for all devices. An output might produce as much as +5 VDC (although the normal range is +2.4 to +3.5 volts), and all input terminals must be capable of accepting that level. Similarly, the properties of a standard TTL input are specified as to current source level in milliamperes. The basic unit used in this specification, then, is the TTL input. With this in mind, the terms *fan-out* and *fan-in* are understandable. The fan-out is an integer number, applied to the output of a device, that tells the designer how many TTL inputs it can drive. If the rating is a "fan-out of 10," the designer knows that the device will drive up to 10 standard TTL input loads. The fan-in is the same type of number, but specifies how many TTL input loads this particular input requires. Most devices have a fan-in of 1 on each input. The designer merely adds up the total fan-in at a point and makes sure that the driving device has a fan-out equal to or greater than this number.

RESISTOR-TRANSISTOR LOGIC

One of the earliest logic families used in IC form was *resistor-transistor logic* (RTL). This family uses type numbers in the μL900 range and 700 range. Two examples are the μL900, which is a popular NAND gate, and the 789, which is a J-K flip-flop.

In most logic families, you get a reasonable picture of typical internal circuitry by examining an inverter, or simple gate, circuit. Figure 4-1 shows an RTL inverter stage. The RTL inverter IC would contain all three components: R1, R2, and Q1.

This circuit should be familiar to all readers, as it is merely a version of the simple common-emitter amplifier circuit using an NPN transistor. Of course, in digital circuits, transistor Q1 is operated in the saturated mode. When a +3.6V potential is applied to the collector, and the ground terminal is connected to the negative side of the power supply, the stage will operate as an inverter. When there is no potential applied to the input, or when the input is grounded, transistor Q1 is cut off. No collector current will flow in Q1, so the output terminal will be at the supply potential (+3.6VDC). Apply a HIGH (+3.6V) potential to the input, however, and Q1 will saturate. This causes the collector to drop to within a few millivolts of ground potential. The output, then, is effectively grounded (a LOW condition). The requirement for

Fig. 4-1. RTL inverter.

inversion is met: A HIGH on the input creates a LOW on the output, and a LOW on the input creates a HIGH on the output.

RTL devices typically operate from a +3.6V power supply, and it was always considered imprudent to operate at higher potentials. A maximum of about +4V is usually considered the rule of thumb. On the positive side of the ledger, however, as long as voltages on a printed circuit board were all less than +4VDC, no combination of shorts or opens could destroy the IC. This is something that cannot be said of the other, more modern, logic families. TTL, for example, is easily destroyed by a slipped screwdriver or careless soldering iron.

RTL is considered to be a current-source family. Note that the output is connected to the +3.6VDC power supply through a resistor, which is similar to the situation in the example of the last section. When the output is HIGH, it will source current to the output through the resistor to the +3.6VDC supply.

DIODE-TRANSISTOR LOGIC

Diode-transistor-logic (DTL) devices are also obsolete, but are still found in equipment now in service. Most DTL devices bear 900-series type numbers.

Figure 4-2 shows a DTL inverter circuit. The integrated circuit device would contain all components shown, such as R1, R2, Q1, and D1. Most of these devices operated at DC potentials of +5 to

+6 volts. The operation is as follows:

☐ When the input is HIGH (+6V), diode D1 is reverse biased. The base of Q1, then, is forward biased, so Q1 will conduct collector current. The output voltage at this time is zero.

☐ When the input is LOW (grounded or at 0V), D1 is forward biased, dropping V_{be} enough to reverse bias Q1. This condition causes the output voltage to go HIGH.

In some DTL circuits, HIGH is +5 to +6 volts, and LOW was 0-volts. In other cases, HIGH and LOW are ±6 volts. DTL operating speeds are approximately the same as RTL.

TRANSISTOR-TRANSISTOR LOGIC

Transistor-transistor logic (TTL) has been the most popular IC logic family for many years—so much so, in fact, as to be called ubiquitous. These devices are given type numbers in the 5400/54000 and 7400/74000 series. The 5400/54000 series devices are able to meet military specifications, while the 7400/74000 devices are the so-called "commercial" grade ICs. Devices with similar type numbers in these series are identical in function, if not specifications. A 5490, for example, is a 7490 in uniform.

TTL devices operate from a +5 VDC power supply. In many TTL devices, the power supply voltage must be maintained within a small range around +5V. The generally given range for all

Fig. 4-2. DTL inverter.

43

TTL devices is +4.75 to +5.2 VDC. Performance is unspecified at below the lower limit, and damage to the device can be expected at above the upper limit.

The increased operating speed (15 to 20 MHz is typical) is achieved at the expense of increased current consumption. The typical TTL device requires 15 to 30 milliamperes. If 22 mA is taken as "average," the current requirements rapidly add up. Fewer than 50 integrated circuits will draw 1A of current from the +5 VDC power supply. Many digital circuits have many more than 50 IC devices! Add to that the fact that optimum circuit performance often requires a power supply capable of delivering a lot more than "just enough" current, and you will see why high-current, 5V power supplies are frequently identified with TTL instruments.

Typical TTL inverter circuits are similar to Figs. 4-3 and 4-4. The device shown in Fig. 4-3 has a typical TTL output, while the circuit in Fig. 4-4 has an open-collector output stage. The open-collector circuit requires an external pull-up resistor, but this feature allows the device to interface directly with lamps, light-emitting diodes, relay coils, and logic circuits operating at non-TTL voltages.

The regular TTL output circuit shown in Fig. 4-3 uses a totem pole stage; transistors Q3 and Q4 form a series pair. Transistor Q2 forms a driver for Q3/Q4, while Q1 forms the input stage. Notice that the input of a TTL device is a transistor *emitter* terminal. The circuit or device, which could be another TTL IC device, that is connected to the input must be able to *sink* the emitter current of

Fig. 4-3. Normal TTL inverter.

Fig. 4-4. TTL inverter with open-collector output.

Q1. It can be said, then, that TTL inputs are current sources, while TTL outputs are current sinks. The operation of Fig. 4-3 follows.

When the Input is LOW:

☐ The emitter of Q1 is grounded, so Q1 is forward-biased.
☐ Resistor R1 and the *c-b* junction of transistor Q2 form the collector load of transistor Q1. Point A, which is where the collector of Q1 and the base of Q2 are connected together, is also LOW.
☐ When point A is LOW, Q2 is cut off, making point B LOW and point C HIGH.
☐ Under this condition, Q3 is turned off and Q4 is turned on, connecting the +5V supply to the output terminal.

When the Input is HIGH:

☐ Transistor Q1 is turned off, so point A is HIGH.
☐ When point A is HIGH, transistor Q2 is turned on, making point B HIGH and point C LOW.
☐ When point C is LOW, Q4 is turned off, disconnecting the +5 VDC supply from the output. Point B is HIGH, so Q3 is turned on. This causes the output terminal to be at ground potential.

An inverter circuit such as Fig. 4-3 needs no external components to operate. But that feature also makes it more diffi-

45

cult to interface the device with anything other than transistors or other TTL devices. In many cases, therefore, an *open-collector* IC is needed. This type of TTL inverter is shown in Fig. 4-4. Notice that the input and driver stages (Q1/Q2) are essentially the same as in the previous case. But only Q3 is used in the output stage, and it has no collector circuit. The collector of Q3 is brought to the outside world through a terminal and must be returned to a positive voltage through external resistor R4. The supply end of R4 must go to +5 VDC in some cases, while it may go to any voltage between +5V and either +15V or +30V in other cases, depending upon type number. An open-collector TTL IC would be used to turn on a lamp, LED, relay, and so forth.

The multiple input gates of the TTL series are similar to the circuits shown previously but will use a multiple emitter input stage (Fig. 4-5). Each TTL input must be a transistor emitter terminal.

EMITTER-COUPLED LOGIC

Emitter-coupled logic (ECL) is a high-speed IC logic family of nonsaturated devices. TTL, on the other hand, operates with the transistors in saturation or cutoff all of the time. Operating frequencies to 120 MHz are common. Some special and very costly devices are able to operate to frequencies in the *gigahertz*, or 1000 MHz, range. Many TTL frequency counters use ECL input stages to divide the input frequency down to TTL ranges, such as 50 to 80 MHz maximum.

Figure 4-6 shows a TTL gate circuit. This particular device is a four-input OR/NOR gate. Note that both inverted (NOR) and noninverted (OR) output terminals are available.

These devices offer optimum AC performance when V_{cc} and V_{ee} are set to ±5 VDC, respectively. The ECL family combines

Fig. 4-5. Multiple inputs are created with extra emitters.

Fig. 4-6. ECL gate.

high speed with moderate power consumption, something usually deemed impossible in the saturated logic families (RTL, DTL, TTL, CMOS, etc.).

HIGH-THRESHOLD LOGIC

The TTL logic devices that have become almost the standards in the electronics industry have only 400 to 500 millivolts of *noise immunity*—the ability to not respond to noise potentials. In many applications, this is not sufficient to eliminate false operation. An industrial site that uses lots of electrically operated machinery is considered electrically noisy. Pulses picked up from the electrical activity will be treated as a valid pulse by TTL and DTL logic elements.

The *high-threshold logic* (HTL), also called *high-noise-immunity logic* (HNIL), offers superior immunity to spurious noise pulses. The first HTL devices were designed to operate from 12V power supplies, while later devices operate from 15V power supplies. CMOS devices are technically high threshold but are not included in this discussion because they are sufficiently different to warrant a separate section. Bipolar HTL devices have approximately 10 times the noise immunity of TTL. The improved noise immunity

47

A TTL

B 12-VOLT HNIL

C 15-VOLT HNIL

Fig. 4-7. Comparisons of noise immunities of HTL and TTL devices.

48

Fig. 4-8. Typical HTL gate.

is derived from three factors: lower circuit impedances, slower operating speeds, and higher threshold potentials.

Additionally, HTL devices usually offer a $\pm 1V$ power supply immunity, as opposed to the $\pm 250mV$ tolerance of TTL. As a result, some circuits using HTL devices do not require regulation of the DC power supply.

Figure 4-7 shows the relative noise immunity of the two HTL families—12V and 15V—with the popular TTL family. Notice that TTL devices have a 400-mV (0.4V) noise immunity band. The 12V HTL/HNIL family has a 3.5V noise immunity band, almost 10 times better. The 15V family, however, does even better, having a 3.2V band at the lower end and 6.5V at the upper end.

Figure 4-8 shows a typical HTL gate. Notice that it is in some respects similar to the DTL circuit design, except for 5.8V Zener diode in series with the input diodes. The Zener diode operates to raise the voltage threshold required to operate the gate. An input signal must be high enough in level to exceed the Zener breakdown potential, or the gate will not respond.

HTL inputs are current sources, so any device driving an HTL input must be able to sink a current equal to approximately 2.1 mA in 12V devices and 2.6 mA in 15V devices.

HTL/HNIL devices can be interfaced with elements of other digital logic families. They can be interfaced directly with RTL/DTL/TTL inputs, *provided* that the V_{cc} supply of the HTL device is the same potential as the RTL/DTL/TTL device. Interfacing with HTL devices operated at other potentials, however, may require some external components. To interface an HTL/HNIL output with a TTL input, an HTL/HNIL device that has an open collector output is usually used. A 3K to 4K pull-up resistor can then be connected between the open-collector output terminal on the HTL/HNIL device to the +5 VDC power supply of the TTL device.

Similarly, TTL open-collector outputs can be interfaced with HTL/HNIL inputs if a pull-up resistor (10K ohm) to the +12V or +15V power supply is provided. A restriction on this latter application, however, is that the TTL open-collector must be a +15V or +30V type, such as a 7406 inverter.

COMPLEMENTARY METAL OXIDE SEMICONDUCTOR LOGIC

Complementary metal oxide semiconductor (CMOS) logic elements are among the most recent on the market, but they are very popular. It is a little premature, however, to write the obituary of the earlier families, especially TTL and ECL. CMOS devices differ from the other families in that they use MOS field-effect transistors instead of bipolar (NPN and PNP) transistors. It is a saturated logic family that is well regarded for its low-current consumption.

It takes two types of MOSFETs to make a CMOS inverter stage: *p-channel* and *n-channel*. The p-channel device will turn on when its gate potential is zero or negative with respect to the source, while the n-channel turns on when the gate is positive with respect to the source.

Figure 4-9 shows a basic CMOS inverter circuit that is designed with an n-channel and a p-channel MOSFET connected in series. The output is taken from their mutual junction—the source end of

Fig. 4-9. CMOS inverter.

one and the drain end of the other. The gates are connected in parallel, so the same potential is applied to both.

When the input signal is LOW (0 volts or some minus potential—both can be used in any given CMOS device), transistor Q1 is turned off and Q2 is turned on. This allows the V_{cc} potential to appear at the output terminal. Only a very small resistor, such as the channel resistance of Q2, is in series with the V_{cc} terminal and the output terminal.

When the input signal goes HIGH, on the other hand, exactly the opposite situation occurs; Q1 is turned on and Q2 is off. In this condition, there is a high resistance between the output terminal and V_{cc} and a low resistance between the output terminal and ground or the minus power supply.

A model of this action is shown in Fig. 4-10, using resistors and the DC supply. The resistances represent the channel resistances of Q1 and Q2 under the two different conditions.

In Fig. 4-10A the input is LOW, in which case Q1 is off and Q2 is on. This makes R_{Q1}, the channel resistance of transistor Q1, HIGH and R_{Q2} LOW. Because of voltage divider action between R_{Q1} and R_{Q2}, the output will be HIGH.

Fig. 4-10. Relative resistances of Q1/Q1 when input is LOW at A, and relative resistances of Q1/Q2 when input is HIGH at B.

The reverse situation occurs, however, when the input is HIGH (see Fig. 4-10B). In this case Q1 is on and Q2 is off. This will reverse the relationship between the respective channel resistances of the two transistors.

The models shown in Fig. 4-10 demonstrate the really unique nature of CMOS devices. These devices will have a low resistance path to V_{cc} in the HIGH input state and a similar path to ground in the LOW input state. The off transistor in any given state will have a high-channel resistance, while the on transistor has a low channel resistance. This situation—a high and a low resistance in series—causes the power supply to always see a high impedance to ground. The only appreciable current drain is when the output stage is in *transition* from one state to another. At those times, the respective on-off roles of the transistors are changing to the opposite state. At sometime during this transition, the channel resistances will be lower than maximum and will be equal. The current at this instant is maximum. During the stable periods the current is in the microampere range.

How do the two most popular IC logic families compare with each other? Both CMOS and TTL devices have their own adherents who will swear *by* them. In truth, though, it must be said that both types of devices have their own virtues and in some areas the proper design selection is made by a coin toss; little or no difference would be noted.

When using the TTL device, be ready to accept the fact that TTL ICs draw a considerable amount of current from the power supply. Additionally, there is the fact that the TTL device must operate from a regulated +5V supply that is negative-grounded. No other supply will do. CMOS, on the other hand, will draw very small amounts of current (microamperes versus milliamperes), and is able to operate with a wide variety of power supply voltages. In many cases, the power supply for CMOS devices need not be regulated. Some designers take advantage that the CMOS device will operate from positive and negative supplies.

The difference in current requirements between CMOS and TTL is, perhaps, best illustrated by two similar gates. A certain TTL NAND gate requires 8 mA when the gate output is HIGH and 22 mA when the output is LOW. A CMOS package that is functionally identical, such as a NAND gate, requires only 15 μA in the HIGH condition and 160 μA in the LOW condition. It is this low power drain feature that makes CMOS very popular, especially in portable equipment. "Credit card" sized pocket calculators and digital wrist watches would not be possible if TTL devices were used! A digital watch uses only a few dozen microamperes in CMOS technology, but you would need to pull a wagon behind you to carry the batteries if both portability and long operation were required of the same design in TTL. In large-scale digital equipment, very heavy power supplies are needed when TTL devices are used. My first (homebuilt) microcomputer, for example, used a lot of TTL and bipolar memory devices (both high current demand), so a 20A power supply at 5V was needed. Some large mainframe computers, the so-called "dinosaurs" (note the microcomputerist's bias), may draw up to 150A to 200A at +5V.

The operating voltage range of the CMOS devices is very wide, while the TTL device must see a regulated +5V supply. Most CMOS devices will operate properly over the range +4.5V to +18V. In many cases, ±15 volts can be used. The minus power supply is used instead of the needed ground connection when monopolar power supplies are used; however, most CMOS devices are designed for optimum performance in the 7V to 12V range. Some complex, special function CMOS ICs will not operate properly at less than +7 VDC.

In contrast to the wide operating range of the CMOS device, we find that the TTL device will not operate well if the power supply voltage deviates very far from the nominal +5V value specified. In fact, many devices are listed as requiring a voltage tightly regulated between +4.7V and +5.2V. It is the experience of many, however, that many circuits will not operate at less than about 4.9V. This is especially true of the more complex special function devices, or where there are a lot of ICs in the circuits. Monostable multivibrators and clock circuits especially tend to become troublesome outside of the +4.9V to +5.2V range.

The high current drain requirements of TTL devices often cause problems in power-supply distribution on printed circuit (PC) boards. The tracks used to carry the +5V and ground return currents will sometimes cause *IR* drops sufficient to reduce the voltage at a chip below the operating range. In addition to this current starvation, there is also the problem of *glitches,* or spurious pulses in the circuit. These raise the dickens with gates, counters, flip-flops, and monostables. A glitch usually consists of a sharp, negative-going pulse on the +5 VDC power supply line. It is caused when a sudden current demand occasioned by turning on a chip reduces the terminal voltage at that chip. Any PC board that contains even a modest number of TTL devices requires bypass capacitors (0.001 μF to 1 μF) generously sprinkled all over the board. There should be at least one 0.001-μF capacitor for each TTL package, and it should be located close to the package +5V line. Disc ceramic and tantalum bypass capacitors are usually specified for this application.

The glitch problem also forces us to use PC boards wherever possible. Regular wiring, increasingly popular on circuit boards due to the availability of low-cost wire-wrapping apparatus, often creates problems due to its inherent inductance. The inductance of a short piece of wire may not seem like much, but it becomes significant when the rise time is very fast, as it often is in TTL circuits, and the current at peak is high. Wire connections should be used only with caution in TTL circuits.

We also find that, in many cases, the noise immunity of the CMOS device is superior to that of the TTL. The TTL device will have but a few hundred millivolts of noise immunity, while the CMOS device has several volts. The reason for this difference is that the CMOS device will not change state unless the input signal passes the $1/2V_{cc}$ point. If a $+12V$ supply is used, for example, the trip point differentiating between HIGH and LOW conditions is 12/2, or 6 volts. The TTL device, on the other hand, is not firmly in the LOW state unless the signal level is less than 800 mV (0.8 volts), and is not in the HIGH state unless the signal has a level over 2.4 volts.

CMOS is a slow logic family while TTL is fast. Almost any garden-variety TTL device will operate to 18 MHz, and most will operate to 20 MHz. Some selected devices or special type numbers operate to 80 MHz. On the other hand, CMOS operates only to 10 MHz or so, and many devices are limited to less than 5 MHz.

Another problem with the CMOS device is that it will be damaged from static electricity potentials that inevitably build up on your body, clothing, tools, instruments and the work bench itself! While most CMOS devices are safe to handle when connected into their circuits, they become exceptionally vulnerable when handled out-of-circuit. This does not, incidentally, give you license to handle the CMOS circuit card carelessly. Damage can and does occur in-circuit; it is merely less probable, not impossible. You will hear some technicians and engineers say that the problem is overstated by both manufacturers and technical writers. After having blown a few terribly expensive devices and caused damage to low-cost devices that were difficult to locate, I can only say, "More power to them." If they want to be the ones who blow it, then let them pay the price. As for me, I will try to use the handling procedures that follow.

The CMOS device is sensitive for the same reason that MOSFET transistors are sensitive: low gate-channel breakdown voltage, which is typically 80 volts. The gate element of a MOSFET device is insulated from the channel by a layer of oxide insulating material that may be as thin as 0.0001 inch. That's 1/10,000! The voltage level of the static electricity on your body is often several hundred volts and may well be several kilovolts. This accounts for the spark and slight shock received when you discharge the static into a grounded object.

Fig. 4-11. MOSFET device showing breach in insulation from static.

Figures 4-11 and 4-12 show the actual mechanism of CMOS/MOSFET failure. When the integrity of the insulating layer is breached, the metal ions of the gate element are drawn into the hole, creating an ohmic path between the channel and the gate. Unfortunately, this problem may not manifest itself immediately. The breach may not be large enough at first to cause an immediate short circuit, but after several weeks of operation there are enough metallic ions in the breach to cause a short. This may account for the disgusting habit some devices have of failing after a few weeks of operation. The cure for this problem is to use special handling procedures and a special working environment in which everything is at the same potential—usually ground.

Some CMOS devices, such as the inverter circuit of Fig. 4-13,

Fig. 4-12. Enlarged view of breached area.

Fig. 4-13. Zener diode gate protection in CMOS gate.

have internal Zener diodes that prevent the breakdown. These diodes will clamp the interelement voltages to some value less than breakdown. These are not perfect, however, and cannot be trusted totally. The B-series CMOS devices, such as 4011B, do have this internal protection.

CMOS devices are shipped from the manufacturer in antistatic containers. They will be in black, plastic bags or on black foam. Some are stored in special containers that are not opaque, but are still antistatic; an ohmmeter reveals that the plastic is conductive. Do not remove CMOS devices from their shipping container or the foam backing until you are ready to use them.

The working environment that you should establish when working on CMOS circuits is a grounded environment. This will allow the static electricity to drain to ground before it can do any damage. One tactic is to use a large layer of aluminum sheet, metallic foil, or even an old cookie sheet as a work surface. This surface must be at ground potential. Similarly, your body and all of your tools and instruments must be grounded. The ground to your body can be an alligator clip attached to your metallic watchband, or it could be a more formal device made from an ID bracelet with a wire soldered to the flat surface.

Caution: It can be *deadly* to keep your body at ground potential if there is any possibility of a high current flowing through it. Two different problems exist. Low-voltage, high-current power supplies will cause a third degree burn that in all probability will require medical treatment if that bracelet contacts the source. Secondly, if that death bracelet touches a 110 VAC source, a lethal AC current will probably flow. Contacting any point of your body with AC will probably be lethal if the environment is directly grounded. Do not ground the worksheet directly. Instead, connect a 1 megohm, or higher, 2-watt resistor between your work surface and actual ground. A grounded work surface could make you a 5000W fuse and is a real widow(er) maker! Using a 1-megohm resistor still allows the surface to be maintained at ground potential but places a limit on the current that can flow. *Please stay alive.*

CMOS Handling Rules

Only a few rules are needed for safe handling of CMOS devices. Note that these same rules generally apply to the Schottky TTL devices discussed in the next section. Failure to follow these rules might cause you to inadvertently build a "silicon to carbon converter."

- ☐ Always handle the CMOS devices, or printed circuit boards containing CMOS devices, in an environment grounded to earth through a 1-megohm, or greater, resistor.
- ☐ Keep CMOS devices or PC boards containing CMOS devices in their antistatic shipping containers until ready for use.
- ☐ When removing the CMOS device or PC board containing CMOS devices from its shipping container, ground yourself first through a 1-megohm resistor. Before actually touching the device or the PC board, touch the container or other conductive shipping surface. This equalizes the pin voltages.
- ☐ Always avoid touching the IC pins.
- ☐ Use only soldering devices that have grounded tips—those that have a three-wire AC power cord.
- ☐ Turn off the power before inserting CMOS devices or PC

Fig. 4-14. Schottky and low-power Schottky inverters.

boards containing CMOS devices into their sockets.

☐ Avoid wearing synthetic fabrics.

☐ Use test equipment that has a three-wire power cord; this indicates that the equipment cabinet is grounded.

☐ Handle PC boards containing CMOS devices only by their edges and avoid touching the exposed card-edge connector.

SCHOTTKY TTL ICS

Schottky TTL devices are a modified TTL family that uses a Schottky barrier diode (SBD) connected between the collector and base of the logic transistors. This technique serves to reduce the base saturation, thereby reducing the charge stored in the transistor. A smaller stored charge means faster switching times.

Figure 4-14 shows an example of two common types of Schottky TTL digital IC logic elements. The low-power Schottky devices are designated by 54LS and 74LS type numbers; i.e., a 74LS121 is a low-power Schottky version of the 74121 TTL device. An "LS" gate will dissipate only 2 mW, with less than 10 nS of propagation delay time.

The regular Schottky devices are designated by 54S and 74S type numbers. They have only 3 nS of propagation delay time, but they require almost 20 mW of power from the DC supply. The "S" and "LS" devices may often be found mixed with regular TTL and are used to reduce system power requirements or enhance speed.

Chapter 5

Logic Gates

After reading this chapter, you'll know the different types of digital logic gates and the different gate ICs available in CMOS and TTL. Additionally, the operation of principal types of gates is covered.

SELF-EVALUATION QUESTIONS

- ☐ What happens to the output if one input of a 7400 gate is grounded?
- ☐ In a NAND gate, the output is _____ when both inputs are HIGH.
- ☐ Both inputs of a two-input XOR gate are HIGH. What is the condition of the output terminal?
- ☐ One input of a two-input NOR gate is HIGH. What is the condition of the output terminal?

Several elementary types of *logic gates* form the basic building blocks out of which even the most complex digital circuits are formed. Even the most complicated computers are merely large scale combinations of a few basic types of logic gate. Some of the digital ICs that will become very familiar to you in the near future as you study this text are merely combinations of different gate circuits in a seemingly monolithic IC form. All complex digital logic functions are made from very elementary gate types.

60

The types of gates which we will discuss are NOT, OR, AND, NAND, NOR, and Exclusive-OR (XOR). Various examples of these gates are found in each of the major IC logic families.

NOT GATES—INVERTERS

Figure 5-1 shows the symbol and truth table for the most elementary type of gate. This gate is called an inverter, complementor, or NOT-gate. This type of stage will invert the input signal; that is to say the output of the inverter will be the opposite of the input. If the input is HIGH, then the output will be LOW. Similarly, if the input is LOW, then the output will be HIGH. Such gates are called *complementors* because, in digital circuits, HIGH and LOW levels are considered to be complements of each other.

We use the NOT terminology because of the symbols used denote input and output. If the input is called A, then the output will be called *not-A* because it will always be the logic state that A is not. The symbol used for not-A is "$\overline{A}$", sometimes called *bar-A*. NOT-A, however, is the preferred expression.

The truth table for a NOT gate is as follows:

A	$\overline{A}$
1	0
0	1

Whenever A is at a logic level 1 (HIGH), then the output $\overline{A}$

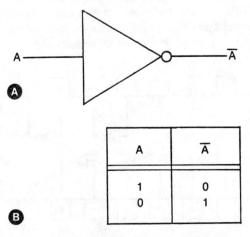

Fig. 5-1. Symbol for inverter at A, and truth table for inverter at B.

will be at logic level Ø(LOW). When A goes to level Ø, then $\overline{A}$ goes to 1.

The symbol for an inverter, or NOT gate, as shown in Fig. 5-1, is a triangle with a circle at the output. Anytime you see a digital logic element in a schematic that has a circle at the output, the output is inverted. Similarly, some multiple-input devices have a circle on one or more inputs. This symbol means that the logic function *for that input only* is inverted.

You will often see a triangle without a circle on the output. This is the schematic representation for *noninverting buffers*. Such devices will produce an output that is identical to the input but is usually capable of delivering more current that ordinary TTL outputs. Buffers are used to isolate one circuit from another, or to increase the drive capability (fan-out) of a circuit. Whenever a logic element must drive a long transmission line or a circuit that is heavily loaded with many inputs, a buffer might be required. Figures 5-2 and 5-3 show popular examples of digital inverters from the TTL and CMOS logic families.

The TTL 7404 device is shown in Fig. 5-2. This chip is called a *hex inverter* because it contains six independent TTL inverters. The positive side of the +5 VDC power supply is applied to pin No. 14, while the ground side is applied to pin No. 7. Although by no means universal, this pinout configuration for the power terminals is very common in digital IC devices. Note that there is no interaction between the inverter sections, so all six may be used independently. The average current drain is 12 to 15 mA.

Three related devices, using the same pinouts, are the 7405, 7406, and 7416. All three of these are *open-collector* chips, meaning that the output transistor requires an external pull-up resistor to the $V_{cc}+$ power supply. The resistor connects the collector of the

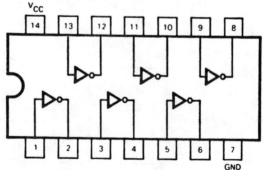

Fig. 5-2. TTL Hex inverter.

transistor, connected internally to the IC output terminal, to its power source. The 7405 is very similar to the 7404, except for the required 2.2K ohm pull-up resistor to the +5V supply. The 7405 draws 12 to 15 mA.

The 7406 and 7416 are open-collector devices, like the 7405, but are able to withstand 30V (7406) and 15V (7416), respectively. Note that the package voltage applied to pin No. 14 *must* remain at +5V; only the output transistor collector can operate at the higher potential. The higher collector voltage is intended to make it possible for the IC to drive an external load.

The 7406 device can sink up to 30 mA, while the 7416 can sink up to 40 mA, through the output transistor collector. This specification means that the 7406 load must have a DC resistance not less than R = (30V)/(0.03A), or 1000 ohms. The 7416 load must be not less than R = (15V)/(0.04A), or 375 ohms. These values refer to the dc resistance between the open-collector output and the V_{cc}+ power supply. Both 7406 and 7416 devices are frequently used as drivers for relay coils, LEDs, and very small incandescent lamps.

Figure 5-3 shows the pinouts for the 4009, 4049, and 4069 CMOS hex inverter chips. The 4009 is now considered to be obsolete for new designs and has been replaced by the 4049 device. The 4009 device uses two V_{cc}+ terminals and will be destroyed if these voltages are applied in the wrong sequence. The voltage applied to pin No. 1 must always be equal to or greater than the voltage applied to pin No. 16. Again, all six inverters may be used independently.

The 4049 device may be used as a level translator between CMOS devices and TTL devices. The voltage applied to pin No. 1 defines the output voltage swing, so if +5 VDC is used, the device becomes TTL-compatible. The 4049 can handle 3.2 mA, so it will drive two standard TTL loads. The input voltages applied to the 4049 inverter sections can swing to +15 VDC, regardless of the V_{cc}+ voltage applied to pin No. 1.

Although some people call the 4069 device a "low-power" version of the 4049, it is not. The 4069 has different pinouts and is not suitable for either direct interfacing with TTL logic or level translation of any other type. The 4069 will only interface directly with other CMOS devices.

The difference between the current drains of TTL and CMOS devices can be seen by comparing the 7404 and 4069 devices. At +5 VDC (used by all TTL devices), the 4069 draws 0.5 mA, as

Fig. 5-3. Hex inverters from the CMOS line.

64

opposed to 15 mA for the 7404; this is a difference of 3000 percent! It is little wonder that designers of battery-powered devices prefer CMOS over TTL.

OR GATES

An OR gate will produce a logical-1 or HIGH output if *any* input is HIGH. Figure 5-4A shows the symbol for an OR gate, Fig. 5-4B shows an equivalent circuit that produces the same action, and Fig. 5-4C shows the truth table for the OR gate.

Let us consider Fig. 5-4B in the light of the definition of an OR gate. Let the voltage at point "C" represent the output of the OR gate, and switches A and B the inputs of the OR gate. The logic is as follows:

Switch Condition	Input logic level	Output Condition
open	0	LOW
closed	1	HIGH

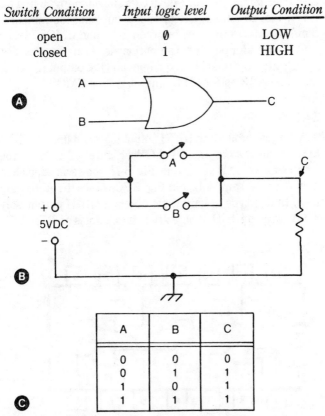

Fig. 5-4. OR gate symbol at A, electrical circuit model for OR gate at B, and truth table for OR gates at C.

By the definition of an OR gate, we would expect C to be HIGH when *either* input is HIGH also (switch A OR switch B is closed), and LOW only when *both* inputs are also LOW (switch A and switch B are open). This is exactly the action found in the circuit of Fig. 5-4B. Voltage C goes to +5 VDC if either switch is closed and will be zero only when both switches are open. This condition is reflected in the truth table shown in Fig. 5-4C: C is HIGH if A OR B is also HIGH.

Examples of TTL and CMOS OR gates are shown in Figs. 5-5 and 5-6, respectively. The type 7432 TTL device is a quad, two-input OR gate. Note that the V_{cc}+ and ground terminals are pins 14 and 7, respectively. This is the same power supply configuration as found on the 7404 (discussed in the previous section). The 7432 requires approximately 20 mA of current per package.

A CMOS example, the 4071 device, is shown in Fig. 5-6. This IC is also a quad, two-input, OR gate. The current requirement is also 0.5 mA, and the difference between TTL and CMOS is seen even more clearly. But power consumption also brings speed. The propagation time—the time between input and the output response that it causes—is 12 ns for TTL, and 80 ns for CMOS.

AND GATES

The AND gate produces a HIGH output if *both* inputs are HIGH and a LOW input if *either* input is LOW. Figure 5-7A shows the circuit symbol for an AND gate, while Fig. 5-7B shows an equivalent circuit that does the same job, and Fig. 5-7C shows the AND gate truth table. In this type of circuit we obtain a HIGH output only if A AND B are also HIGH. There are few examples of TTL AND

Fig. 5-5. TTL OR gate.

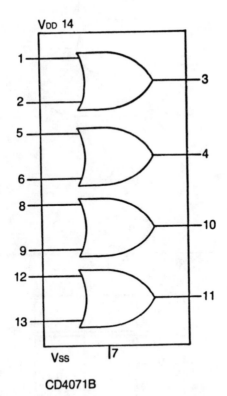

Vss 7

CD4071B

Fig. 5-6. CMOS OR gate type 4071.

gates, but in the CMOS line we find the 4073, 4081, and 4082 devices. The 4073 is a triple, three-input AND gate. The 4081 is a quad, two-input AND gate, and the 4082 a dual, four-input device.

NAND GATES

A NAND gate can be viewed as an AND gate with an inverted output; if fact, you can simulate the NAND gate by connecting the input of an inverter (NOT gate) to the output of an AND gate. The designation "NAND" is derived from *NOT-AND.* We sometimes see the NAND expression written in the form $\overline{AND}$, $\overline{AXB}$, or $\overline{A.B}$.

Figure 5-8 shows the NAND gate. Figure 5-8A shows the circuit symbol used in schematic diagrams. Figure 5-8B is an equivalent circuit that performs the same job, and Fig. 5-8C is the NAND gate truth table. The rules of operation for the NAND gate are as follows:

☐ A LOW on *either* input creates a HIGH output.

67

A	B	C
0	0	0
0	1	0
1	0	0
1	1	1

Fig. 5-7. AND gate symbol at A, electrical circuit model for an AND gate at B, and truth table for an AND gate at C.

☐ A HIGH on *both* inputs is required for a LOW output.

We can see this same action in the equivalent circuit shown in Fig. 5-8B. Recall out switch logic protocol: Switch open is a LOW on the input, and a switch closed in a HIGH on the input.

The voltage at point "C" will be HIGH if either switch A or switch B is open (either input is LOW). But if both switches are closed (both inputs are HIGH), the load is shorted out and point "C" is LOW.

The NAND gate is one of the most popular gates used in digital equipment. An example of a quad, two-input, TTL NAND gate is shown in Fig. 5-9, the popular 7400 device. It is literally possible to construct circuits to replace any of the other gates, using only 7400 devices, although such would be an unnecessary and uneconomic course of action.

A	B	C
0	0	1
0	1	1
1	0	1
1	1	0

Fig. 5-8. NAND gate symbol at A, electrical circuit model for a NAND gate at B, and truth table for a NAND gate.

Fig. 5-9. 7400 TTL NAND gate.

69

Recall that TTL is a current-sinking logic family, meaning that the output will sink, or pass to ground, current. The inputs of a TTL device form a 1.6-mA current source. Any TTL input is considered HIGH when greater than +2.4V is applied, and must be held to less than 0.8V when LOW. This requirement means that any ground connection must have a resistance less than R = (0.8V)/(0.0018 A), or 440 ohms. If a higher ground resistance is encountered, or if a voltage between 0.8V and 2.4V is applied, then operation is undefined—and all bets as to operation are off! The 7400 devices requires approximately 12 mA of current per package.

A CMOS NAND gate IC, the 4011, is shown in Fig. 5-10. This device is also a quad, two-input, NAND gate. The 4011 requires only 0.4 mA per package, again a substantial savings over TTL. Various combination circuits using NAND gates will be discussed in subsequent chapters.

Fig. 5-10. CMOS NAND gate.

Fig. 5-11. Symbol for a NOR gate at A, electrical circuit model for a NOR gate at B, and truth table for a NOR gate at C.

NOR GATES

A NOR gate is the same as an OR gate with an inverted output. The NOR designation is derived from "NOT-OR," and the NOR functions can be symbolized by $\overline{A + B} = C$. The circuit diagram symbol for the NOR gate is shown in Fig. 5-11A, while the equivalent circuit is shown in Fig. 5-11B. A truth table for the NOR gate function is shown in Fig. 5-11C. The rules governing the operation of the NOR gate are:

☐ A HIGH on *either* input produces a LOW output.

71

□ If *both* inputs are LOW, then the output is HIGH.

We can see this action in the circuit of Fig. 5-11B. If either switch A or switch B is closed, then output C will be LOW. But when both switches are open, then output C is HIGH.

Figure 5-12 shows an example of a popular TTL 7402 NOR gate. This device contains four two-input gates that may be used independently of each other. The package draws about 12 mA.

EXCLUSIVE-OR (XOR) GATES

The exclusive-OR, or XOR as it is designated, is shown in Fig. 5-13. Examine the truth table for the XOR gate below:

A	B	C
0	0	0
1	0	1
0	1	1
1	1	0

From this truth table we may infer the rules for operation governing the exclusive-OR gate:

□ The input is LOW if *both* inputs are LOW or if *both* inputs are HIGH.
□ The output is HIGH if either input is HIGH.

In other words, a HIGH on either input *alone* produces a HIGH output, but a HIGH on both inputs simultaneously produces a LOW output. Any time we find all inputs of an XOR gate in the same condition (HIGH or LOW), then the output will be LOW.

Fig. 5-12. 7402 TTL NOR gate.

A	B	C
0	0	0
1	0	1
0	1	1
1	1	0

Fig. 5-13. Symbol for an exclusive-OR gate at A, and truth table for an exclusive-OR gate at B.

SUMMARY OF GATE ACTIONS

In this chapter we have discussed five basic gates (AND, OR, NAND, NOR, and XOR), and the inverter (sometimes called a NOT gate). From these basic forms of gate, *all* digital functions can be constructed; in fact, even the largest special function LSI (large-scale integration) digital ICs contain little more than a complex interconnection of these six basic functions: AND, OR, NAND, NOR, XOR, and NOT. If we want to split hairs even further, we could say that only AND, OR, XOR, and NOT are needed, because the NOR and NAND functions are NOT-OR and NOT-AND, respectively.

Figures 5-14 through 5-19 show a series of waveform timing diagrams that summarize the actions of these basic gates. Such diagrams are frequently used in digital electronics as a graphic means of showing dynamic circuit action. In fact, in many cases the timing diagram is the only reliable way to visualize circuit action.

In the timing diagrams, we have followed the same denotation as before: A and B are inputs, while C is the output. Note that these diagrams are for two-input devices (except, of course, for the inverter), but the basic concepts can be extended to devices with three or more inputs as well.

Fig. 5-14. AND gate waveforms.

In Fig. 5-14, we see the operation of the AND gate. In this case, a train of squarewaves is applied to input A, while a single pulse is applied to input B. Note that output C remains LOW until input B goes HIGH. This is due to the requirement that both inputs be HIGH before the output can be HIGH. Once Input B has gone HIGH, because of the same rule, the output will follow input A. To emphasize again, the rules for AND gates are:

☐ The output will be LOW if either input is LOW.
☐ The output is HIGH only if both inputs are HIGH.

A timing diagram for the NAND gate is shown in Fig. 5-15. Recall that a NAND gate is an AND gate with an inverted output. The NAND gate circuit symbol shows this relationship because it is an AND gate symbol with a small circle at the output terminal.

The NAND gate timing diagram uses input condition similar to the previous case for the purpose of comparing their operations. A squarewave is applied to input A, while a single pulse is applied to input B. If either input is LOW, then the output will be HIGH. Therefore, the output is HIGH until both A and B are HIGH. When B goes HIGH, the output will follow A, but is inverted, or 180

Fig. 5-15. NAND gate waveforms.

74

Fig. 5-16. OR gate waveforms.

degrees out of phase. To reemphasize, the rules for NAND gates are:

☐ The output is HIGH if either input is LOW.
☐ The output will be LOW only if both inputs are HIGH.

The timing diagram for an OR gate is shown in Fig. 5-16. Recall that an OR gate will produce an output if either input is HIGH. This action is shown in the output waveform of Fig. 5-16. The output waveform would also be HIGH if both inputs were HIGH. To reemphasize the action of an OR gate, the rules are as follows:

☐ The output is HIGH if either input, or both inputs, are HIGH.
☐ The output is LOW only when both inputs are also LOW.

Fig. 5-17. NOR gate waveforms.

75

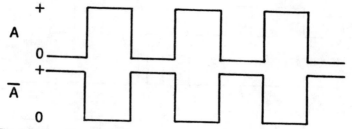

Fig. 5-18. Inverter waveforms.

The timing diagram for the NOR gate is shown in Fig. 5-17. Recall that the output of a NOR gate will be LOW if either input is HIGH. To reemphasize, the rules for the operation of a NOR gate are:

☐ The output will be LOW if either or both inputs are HIGH.
☐ The output will be HIGH only if both inputs are LOW.

The timing diagram for an inverter is shown in Fig. 5-18. Notice that the output A̅ is merely the inverse of the input; that is, it is exactly 180 degrees out of phase with the input. Although this seems like a trivial case, it is none the less important.

The exclusive-OR gate timing diagram is shown in Fig. 5-19. Recall that the XOR gate will produce a HIGH if either input is HIGH, but not if both inputs are HIGH simultaneously. If both inputs are HIGH, or if both inputs are LOW, then the output will be LOW. This action can be seen in Fig. 5-19. Input A sees a train of squarewaves, and we have labeled the different sections 1, 2, 3, . . . etc. to more closely study the effects. At time 1, both inputs

Fig. 5-19. XOR waveforms.

are LOW, so the output C is also LOW. But at time 2, input A is HIGH and input B is LOW. This will make output C HIGH until the beginning of time 3. At that time, both inputs are again LOW. Note in time period 4 that both inputs A and B are HIGH, so the output is LOW. Similarly, the output is LOW during the time period 8. To reemphasize, the rules for the exclusive-OR (XOR) gate are as follows:

☐ The output is LOW any time that both inputs are in the same logic condition—HIGH or LOW.

☐ The output is HIGH if either input is HIGH, except when both inputs are simultaneously LOW.

You may have noted in some of the descriptions in this chapter that one must be careful in using the terms AND/OR and NAND/NOR. In the NAND gate description, for example, the output is HIGH if either A OR B is LOW. But isn't this NOR gate talk? No, not if we are talking about positive logic. Most of the designations are based on positive logic. But when you talk about negative logic, in which logical-1 is zero volts, and logical-0 is V+, the definitions reverse. A positive logic NAND gate is a negative logic NOR gate. Similarly, a positive logic AND gate is a negative logic OR gate. This is why some older catalogues, published before the "standard" of positive logic was firmly established, listed the 7400 as a NAND/NOR gate; it all depends upon your point of view.

EXPERIMENTS

The experiments to follow give you the chance to examine the behavior of the various types of gate. A dual trace oscilloscope is useful in these experiments, but one can "make do" without such an instrument. You can, for example, use the logic level detector circuits shown in Appendix A, or the multichannel electronic oscilloscope switch shown in Appendix B. The oscilloscope switch will allow you to use a single trace oscilloscope to view two traces. The essential thing is not whether you know how to use an oscilloscope, but that you learn to recognize logic levels.

Experiment 5-1

A type 7404 inverter is used in this experiment. The purpose of the experiment is to show the operation of the inverter circuit—the NOT gate.

Fig. 5-20. Experiment circuit.

Without an oscilloscope, connect the circuit as shown in Fig. 5-20. Use the LED logic level indicators (Appendix A). These will be *on* for HIGH and *off* for LOW.

With a two-channel oscilloscope, connect the circuit in Fig. 5-21. Position the trace carrying the 7404 input on the top of the CRT screen, and the trace showing the 7404 output on the lower portion of the CRT screen.

When switch S1 is connected to the "1" position, the input of the 7404 is HIGH. When S1 is connected to the "0" position, the input of the 7404 is grounded and is LOW.

☐ Place S1 in the "1" position and note the LEDs (or oscilloscope trace). The input is _____, so the output is _____.

☐ Place S1 in the "0" position, and note the LEDs (or oscilloscope trace). The input is _____, so the output is _____.

Experiment 5-2

Perform the hook-up for Experiment 5-1, but replace switch S1 with the transistor switch circuit shown in Fig. 5-20. Disconnect S1 and remove it from the circuit.

Connect a 1-Hz squarewave generator to the input of the transistor switch circuit. Note the LEDs or the oscilloscope trace; the LEDs should be blinking back and forth as the squarewave goes HIGH and LOW.

If you are using a two-channel oscilloscope, you might wish to examine the input and output waveforms at a higher frequency—over 200 hertz—and compare them with Fig. 5-18. Most oscilloscopes will be easier to trigger or synchronize at the higher frequency.

Experiment 5-3

This experiment demonstrates the 7400 NAND gate. Connect the circuit of Fig. 5-22. If you have an oscilloscope or digital logic analyzer with three or more channels, use it instead of the LED level indicators.

☐ Open both S1 and S2 (place both switches in the "1" position).

Fig. 5-21. Experiment circuit.

Fig. 5-22. Experiment circuit.

Fig. 5-23. Pinouts for 7402 at A and 7486 at B for use in experiments.

☐ The output level is (HIGH/LOW) _____.
☐ Close S1.
☐ The output is (HIGH/LOW) _____.
☐ Open switch S1 and close switch S2.
☐ The output is (HIGH/LOW) _____.
☐ Close S1 and S2.
☐ The output is (HIGH/LOW) _____.

Experiment 5-4

The 7400 TTL NAND gate IC is used to operate as an on-gate for a pulse train. The 7400 IC can be used—and often is—to control the flow of clock pulses or data into the circuit. Replace switch S2 in Fig. 5-22 with a transistor switch such as Q1 in Fig. 5-20.

Without an oscilloscope, connect a 1-Hz squarewave source to the input of the transistor switch. With an oscilloscope, connect a squarewave source to the input of the transistor switch. Use a frequency over 200 Hz for best oscilloscope display.

☐ Set S1 to the "0" position
☐ The output is (HIGH/LOW) _____
☐ Set S1 to the "1" position
☐ The output is (HIGH/LOW) _____

The output should pulse, or follow the input signal (inverted), *only* when S1 is set to "1."

Experiment 5-6

This experiment uses the exclusive-OR (XOR) gate. A TTL 7486 device is selected.

Perform experiments 5-3 through 5-5 using a 7486 XOR gate instead of the gates called for originally. The pinouts for the 7486 are shown in Fig. 5-23B. Compare the results in each case with the results obtained previously in each experiment and the rules for the XOR gate.

Chapter 6

Arithmetic Circuits

You're going to learn here about the half-adder, adder, and subtractor circuits. These are known as arithmetic circuits. Also, you will find out about the different types of arithmetic logic chips available to perform these functions.

SELF-EVALUATION QUESTIONS

- ☐ Describe the principal difference between half-adder and adder circuits.
- ☐ What is the function performed by the TTL 7483?
- ☐ What is the function performed by the TTL 7485?
- ☐ Draw a circuit diagram for a half-adder circuit.

ADDER CIRCUITS

There are two basic forms of adder circuit. These are called the *half-adder* and *full-adder* (or simply *adder*). Of these, several different types exist. But before examining the various types of circuits, consider the process of binary addition to find out the requirements placed on any circuit that purports to be a binary adder.

The rules for binary addition are as follows:

$$0 + 0 = 0$$
$$0 + 1 = 1$$
$$1 + 0 = 1$$
$$1 + 1 = 0 \text{ plus carry } 1$$

Examine these rules and compare them with the truth table for an exclusive-OR gate. An XOR gate obeys the following rules:

☐ If both inputs are zero, then the output is zero.
☐ If both inputs are one, then the output is zero.
☐ A one output is created if either, *but not both*, inputs are one.

The truth table for the XOR gate, then, is

A	B	Output
0	0	0
0	1	1
1	0	1
1	1	0

Note that the sole difference between the truth table for the XOR gate and the rules for binary addition is that, in binary addition, a *carry-one* output is created if both A and B are 1. We can therefore create an addition circuit with an XOR gate and a means for generating the carry-one output.

Figure 6-1 shows a circuit for a *half-adder*. A half-adder is a circuit that generates two outputs: *sum* (S = A + B) and *carry* (C = A × B) (*read:* S = A XOR B and C = A AND B). The exclusive-OR gate generates the sum output, while an AND gate generates the carry output. In many cases, the logic symbol shown in Fig. 6-2 is used to denote the half-adder in logic diagrams, block diagrams, or flow charts.

A full adder, or simply adder as it is usually called, is a circuit that will accept as a valid input the carry output from a lower order stage (another adder). Consider again the process of binary addition. Add two binary numbers:

83

Fig. 6-1. Half-adder circuit.

In the first step, we added the two least significant digits. In this case they are both ones, so the result is zero, and a carry output 1 is generated. This carry one is added to the next most significant column. In the second step, we add a 0 and a 1 to obtain a 1 result. Next we add the carry one from the previous step, so this is 1 + 1. The result (written down below the line) is a 0, with a carry-1 generated. The result of this operation is a final result of 00 and a carry 1 output. The carry 1 output can be ignored if there is no more significant digits.

The first operation in the above problem could be performed by a half-adder circuit because the least significant digit does not receive any carry bits from a lower order stage (there are no preceding stages to generate any carries). All subsequent stages, however, must be full adder circuits so that they can accommodate carries from previous stages. Otherwise, an erroneous output result could be obtained.

Figure 6-3 shows the logic block diagram for a full-adder, which

Fig. 6-2. Half-adder circuit symbol.

84

Fig. 6-3. Full-adder circuit.

by definition will:

□ Add A and B (S = A XOR B).
□ Account for any carry-in C from lower order stages.
□ Produce a carry-out 1, C_o, if needed.

The circuit of Fig. 6-3 consists of two half-adders and an OR gate to generate the carry output. Half-adder HA1 adds together the A and B inputs using the XOR function. This circuit will produce a carry-out 1 if needed. Half-adder HA2 adds together the results of HA1 (S′) to produce the final output: S = S′ XOR C_i. A carry-out may also be generated by this stage. A commonly used logic symbol for the full adder is shown in Fig. 6-4.

SUBTRACTOR CIRCUITS

Addition and subtraction are very similar arithmetic operations. In fact, it is proper to view subtraction as the addition of a positive and a negative number. It is not surprising, then, to find certain similarities between adders and subtractors used in digital circuits; in fact, the principal difference is the capability to account for a borrow-1 from the next higher order digit.

Figure 6-5 shows a simple half-subtractor circuit which produces an output difference *d* of A XOR B, and a borrow *b* of $\overline{A}$ AND B. The logic symbol for a half-subtractor is shown in Fig. 6-6.

Fig. 6-4. Full-adder circuit symbol.

Fig. 6-5. Half-subtractor circuit.

Fig. 6-6. Half-subtractor symbol.

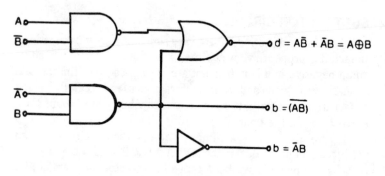

Fig. 6-7. TTL subtractor made from standard NAND, NOR and inverters.

An example of a half-subtractor circuit made entirely from standard TTL NAND, NOR, and NOT gates is shown in Fig. 6-7. This type of circuit is often found in digital equipment because the low cost of TTL IC devices makes it economical.

A full-subtractor must not only find the difference between the two inputs, and create the borrow, but must also account for borrows from other stages. Figure 6-8 shows the full-subtractor

86

Fig. 6-8. Full-subtractor.

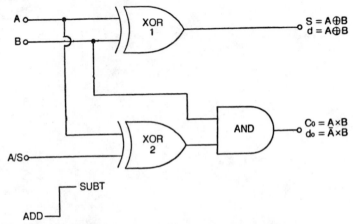

Fig. 6-9. Adder/subtractor made from XOR and AND gates.

circuit, in which two half-subtractors and an OR gate are used.

A universal adder/subtractor circuit is shown in Fig. 6-9. Recall that both adders and subtractors produce an output of A XOR B. The principal difference is that adders produce a *carry* signal

Table 6-1. Half-Adder Truth Table.

INPUTS		OUTPUTS	
A	B	S	C
0	0	0	0
0	1	1	0
1	0	1	0
1	1	0	1

(C_o = A AND B), while subtractors produce a *borrow* signal (b_o = $\overline{A}$ AND B).

In Fig. 6-9, the output of XOR1 will be A XOR B for both modes: addition or subtraction. The input of XOR2 marked "A/S" is the mode control terminal and selects whether the addition or subtraction is used. If A/S is LOW, then the output of XOR2 will be LOW if *A* is HIGH. The output of the AND gate, then, will be HIGH when *A* is HIGH AND *B* is HIGH (A XB). This is a *carry*, so the circuit operates as an adder when A/S is LOW.

When A/S is HIGH, the output of XOR2 will be HIGH when *A* is LOW. We can, therefore, state that the output of the NAND gate is $\overline{A} \times$ B, which is the definition of a *borrow* function. The circuit of Fig. 6-9, then, operates as a subtractor when A/S is HIGH.

Chapter 7

Flip-Flops:

Clocked and Unclocked

The objective of this chapter is for you to learn the operation of clocked and unclocked flip-flops. You will also become familiar with the various TTL/CMOS flip-flops available.

SELF-EVALUATION QUESTIONS

☐ An R-S flip-flop made with _____ gates produces a Q = 1 output when the S input is momentarily brought LOW.

☐ An R-S flip-flop made with _____ gates produces a Q = 1 output when the S input is made momentarily HIGH.

☐ A _____ flip-flop transfers its input data to the output on the arrival of clock pulses.

☐ A J-K flip-flop goes to a state of Q = 1 if the _____ input is made HIGH.

All computer and digital logic functions can be made from combinations of AND, OR, NAND, NOR, and NOT gates; in fact, even the NAND and NOR functions are merely combinations of NOT gates with AND and OR gates, respectively. All of the gate functions are transient circuits; that is, they are incapable of *storing* information, even for a short period of time.

Flip-flops are circuits that will store single-bits of information. Most solid-state computer memories are arrays of flip-flops

Fig. 7-1. NOR gate version of an R-S flip-flop.

organized in a manner that allows storage of the digital words of a computer.

Most flip-flops are *bistable* circuits; that is, there are two stable output states. The flip-flop is not particularly concerned which state is in existence at any given time; it is happy in either state. If the output is labeled Q, we find that Q can be either HIGH or LOW and will be stable when it is HIGH or LOW.

R-S FLIP-FLOPS

One of the simplest flip-flops is the *reset-set* (R-S) flip-flop. These circuits can be made from either NAND or NOR gates, although the performance characteristics are different for the two different types. The NOR gate implementation is shown in Fig. 7-1, while an occasionally used circuit symbol is shown in Fig. 7-2.

Note that the circuit has two outputs, labeled Q and $\overline{Q}$ (read "not Q"). These outputs are complementary, meaning that one will be HIGH while the other will be LOW. If Q is HIGH, then not-Q must be LOW. Similarly, when not-Q is HIGH, Q must be LOW.

In the NOR gate R-S flip-flop, the output changes state when an appropriate input is momentarily brought HIGH. Only a brief pulse at the input is needed to effect the change. The rules governing the operation of the NOR-gate R-S flip-flop are

Fig. 7-2. Circuit symbol for the R-S flip-flop shown in Fig. 7-1.

Table 7-1. Truth Table for an R-S Flip-Flop.

S	R	Q	$\overline{Q}$
		(no change)	
0	0		
1	0	1	0
0	1	0	1
1	1	(disallowed)	

summarized in the truth table shown in Table 7-1. These rules follow.

☐ When both *S* and *R* input are LOW, no change occurs in the output.

☐ If both *S* and *R* are simultaneously brought HIGH, this is a *disallowed state*. This input condition is to be avoided.

☐ Bringing the *S* input momentarily HIGH will *set* the output; i.e., make Q HIGH and not-Q LOW.

☐ Bringing the *R* input momentarily HIGH forces the flip-flop to *reset*; i.e., make Q LOW and not-Q HIGH.

The basic NOR gate R-S flip-flop can be constructed from either TTL or CMOS NOR gate ICs. In the case shown in the example, TTL type 7402 NOR gates have been used to make the flip-flop.

The NAND gate R-S flip-flop uses inverted logic; i.e., the output state changes are caused by bringing the appropriate input LOW.

Fig. 7-3. NAND gate version of an R-S flip-flop.

Fig. 7-4. Circuit symbol of the R-S flip-flop shown in Fig. 7-2.

An example of a NAND gate R-S flip-flop is shown in Figs. 7-3 and 7-4.

Because the inputs are active-LOW, this circuit is sometimes called the reset-set (or R-S) flip-flop. In this text, however, it is more convenient to simply refer to both types as R-S flip-flops, and then be sure that the NAND/NOR distinction is made in the text. The example of Fig. 7-3 is implemented using the type 7400 TTL NAND gate.

The output conditions are governed by the following rules and summarized in the truth table shown in Table 7-2.

☐ If the $\overline{S}$ and $\overline{R}$ inputs are both simultaneously LOW, we have a *disallowed state*. Such states must be avoided.

☐ If both S and R inputs are made simultaneously HIGH, then no change in the output state will occur.

☐ If the S input is momentarily brought LOW while the R remains HIGH, then the Q output goes to the HIGH state and not-Q is LOW.

☐ If the R input is momentarily brought LOW and the S input remains HIGH, then the Q output goes LOW and the not-Q goes HIGH.

The R-S flip-flop is used frequently in applications where a pulse applied to the S or R input sets a condition and then a subsequent pulse is used to reset the circuit to its initial conditions.

CLOCKED R-S FLIP-FLOPS

The operation of the two previous R-S flip-flops was uncon-

Table 7-2. Truth Table for the NAND Gate R-S Flip-Flop.

$\overline{S}$	$\overline{R}$	Q	$\overline{Q}$
0	0	(disallowed)	
1	0	0	1
0	1	1	0
1	1	(no change)	

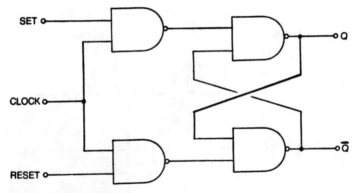

Fig. 7-5. Clocked R-S flip-flop.

ditional. The output state changed immediately when an appropriate input pulse was received. Such circuits are said to operate *asynchronously*.

The clocked R-S flip-flop of Fig. 7-5 is able to operate in a synchronous manner; i.e., the output will change state only when the input pulse coincides with a *clock* pulse. This behavior is obtained by adding a pair of NAND gates to the NAND-type R-S flip-flop circuit.

A commonly used circuit symbol for the clocked R-S flip-flop is shown in Fig. 7-6. This circuit is sometimes called an RST flip-flop.

An example of a timing diagram for an RST flip-flop is shown in Fig. 7-7. The clock produces a fixed-frequency chain of squarewaves at the C input. Note that the Q output does not change state immediately when the S input pulse goes HIGH. The output waits until the clock input is also HIGH. Similarly, the output is not reset until the clock pulse and the R input pulse are coincident.

MASTER-SLAVE FLIP-FLOPS

It is often difficult, or impossible, to transfer data through a circuit in an orderly manner. We often get into high-speed electronic analogies of the old-fashioned "relay race" problem. Recall that type of problem in circuits where two relays are supposed to close

Fig. 7-6. Circuit symbol of the clocked R-S flip-flop.

Fig. 7-7. Typical operation of a clocked R-S flip-flop.

simultaneously? If one relay is a little sluggish or the other is a little faster than usual, then there will be a brief instant where one is open and the other is still closed. This condition often produces unpredictable or disastrous results. The same action exists in digital circuits, and is caused by device propagation time.

Note that two RST flip-flops and an inverter can be used to synchronize a data transfer. The circuit, which is shown in Fig. 7-8, is called a *master-slave flip-flop*.

In Fig. 7-8, the inputs to FF1 are the inputs to the circuit as a whole. The outputs for the overall circuit are the outputs of FF2. Also, the clock affects FF2 directly but must pass through an

Fig. 7-8. Master-slave flip-flop.

94

inverter before it can affect FF1. Both FFs only become active when their respective clock inputs are HIGH.

A timing diagram for this circuit is shown in Fig. 7-9. Recall from above that FF1 is enabled when the clock pulse is LOW, and FF2 is enabled when the clock pulse is HIGH.

When the pulse is applied to the S input, nothing happens at Q1 (the Q output of FF1) until the clock pulse drops LOW. At that time Q1 snaps HIGH, thereby making the S input of FF2 HIGH. But at this time the clock pulse is LOW, so no change occurs at the FF2 output terminal. When the next clock pulse arrives, however, the FF2 clock input goes HIGH (Q1 is still HIGH), so the Q output of FF2 goes HIGH.

Similarly, the reset pulse arrives when the clock is LOW, so the Q1 output immediately drops LOW. The Q output of FF2, however, remains in the HIGH condition until the HIGH transition of the next clock pulse.

FF1 is considered the master flip-flop, while FF2 is the slave. The action at FF1 is given time to settle before the changes can be reflected at the output of FF2. This provides an orderly transfer of data between input and output.

Fig. 7-9. Typical operation of a master-slave flip-flop.

95

Fig. 7-10. Making a type-D flip-flop from an R-S flip-flop.

TYPE-D FLIP-FLOPS

A type-D flip-flop is a modified RST flip-flop that has only one input (labeled the *D* or *data* input). See Fig. 7-10. The type-D flip-flop will transfer data from the D input to the Q output *only* when the clock terminal is HIGH. The following are the rules governing the operation of the type-D flip-flops:

☐ When the clock input goes HIGH, the data present on the D input is transferred to the Q output.

☐ If the clock input *remains* HIGH, the Q output will *follow* changes in the data present at the input.

☐ If the clock remains LOW, then the Q output will *retain* the data that was present on the D input at the instant the clock dropped LOW.

Because of the behavior presented in the above rules, the type-D flip-flop is sometimes called a *data latch*, or simply *latch*. The circuit symbol is shown in Fig. 7-11.

We can see these rules more graphically in Figs. 7-12 through 7-14. Consider Fig. 7-12 first. Recall that the data on the D input will be transferred to the Q output only when the clock terminal

Fig. 7-11. Circuit symbol of a type-D flip-flop.

96

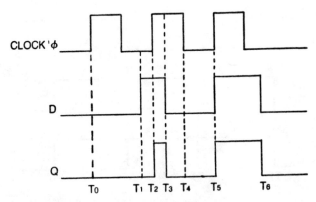

CLOCK 'φ

D

Q

T0   T1 T2 T3 T4   T5   T6

Fig. 7-12. Typical operation of a type-D flip-flop.

is HIGH. At time t_o in Fig. 7-12, the clock goes HIGH, the D input is LOW, though, so the Q output is LOW also.

At time t_1, the D input goes HIGH but cannot affect the Q output because the clock is LOW. The clock goes HIGH at t_2, so the output will also go HIGH. The output pulse exists only for the interval $t_3 - t_2$, because the HIGH conditions on the D input only coincide in that interval.

At time t_5, both the D and the clock inputs go HIGH, but note that the D line remains HIGH even after the clock pulse disappears. By the third rule, then, the output must remain HIGH after the clock pulse passes.

Another situation is shown in Fig. 7-13. In this case, the clock line goes HIGH and remains HIGH. The Q output, therefore, is in an unlatched condition, so will follow the data at the D input. Because the D input data is a squarewave, the output data will also be a squarewave.

C

D

Q

Fig. 7-13. Operation with clock HIGH.

97

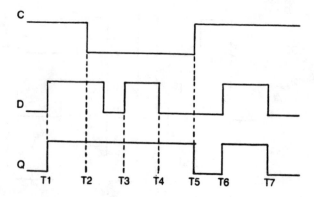

Fig. 7-14. Another operation.

Still another condition is shown in Fig. 7-14. Again the D input data is a squarewave, but the clock is not permanently HIGH in this example. At time t_1, both the clock and the D input are HIGH, so the Q output is also HIGH. At time t_2, the clock drops LOW, but since D is HIGH at that instant, the output will remain in the HIGH condition. Note that the D input can change at will, without affecting the output, as long as the clock input remains LOW. But at the time t_5, the clock goes HIGH, and since the D line is LOW, the output goes LOW also. At time t_6, however, the D line goes HIGH, while the clock is still HIGH, so the output goes HIGH.

Examples of Type-D Flip-Flops

Only rarely will modern circuit designers use individual logic gates to make a flip-flop of any variety. The only common example is the R-S flip-flop. Even those, however, are commonly available in the form of standard CMOS chips, where they had not been in TTL. There are too many good integrated circuit flip-flops on the market for anyone to seriously consider building their own. Figures 7-15 through 7-18 show two TTL and one CMOS version.

A TTL dual type-D flip-flop is shown in Fig. 7-15. The logic diagram is shown in Fig. 7-16. This particular type-D flip-flop has both *set* and *clear* inputs, in addition to the normal clock and D inputs. The 7474 device will obey the following rules of operation:

☐ Data on the D input is transferred to the Q output only on positive-going transitions of the clock pulse.
☐ No changes in the data are reflected as output changes

98

Fig. 7-15. TTL dual type-D flip-flop.

unless the clock is *going* HIGH. The device is, therefore, an edge-triggered flip-flop.
☐ If the *set* input is grounded, then Q immediately goes HIGH and not-Q goes LOW.

Fig. 7-16. TTL dual type-D flip-flop.

- If the *clear* input is grounded, then Q goes LOW and not-Q goes HIGH.
- Set and clear inputs must not be simultaneously grounded. If these inputs are not being used, then they should be tied HIGH to +5 VDC.
- The two-flip-flops in the 7474 IC are completely independent of each other, except for sharing common power supply and ground terminals.

The TTL 7475 device shown in Fig. 7-17 is a special case of the type-D flip-flop called a *quad-latch*. This IC device is level sensitive, so the output will follow changes as long as the clock terminal is HIGH. This is more like the behavior expected of traditional type-D flip-flops, as given earlier in this section.

The 7475 is used primarily to hold four bits of data; i.e., one bit in each flip-flop. The pinouts for the 7475 are shown in Fig. 7-17. The clock inputs are labeled *enable* inputs. When the enable inputs are HIGH, then, data on the D inputs are transferred to the Q outputs. The 7475 is arranged in a 2 × 2 format, meaning that the enable inputs of two flip-flops are tied together and are brought out to a package pin. The enable inputs to the other two remaining flip-flops are also tied together and are brought out to another package terminal.

One common place to find 7475s is in digital counter applications, where they are used to hold the BCD data being displayed while the counter chip, which is often a 7490 device, updates the data. The display can be updated when the count is completed by momentarily bringing the enable terminals HIGH. A single 7475 will store all four bits required for a single BCD display decoder.

54LS/75LS75

Fig. 7-17. TTL quad latch is a kind of type-D flip-flop.

100

Fig. 7-18. CMOS dual type-D flip-flop.

A device similar to the 7475 is the TTL IC type 74100. This 24-pin IC contains a pair of four-bit latches and therefore has the ability to store up to eight bits of digital data, a fact that is not lost on microcomputer circuit designers.

A CMOS type-D flip-flop is the CD4013 or, simply 4013, device shown in Fig. 7-18. Like the 7474 device from the TTL line, the CMOS CD4013 is a dual type-D flip-flop. It can be used in either direct or clock modes, i.e., it has set and clear direct inputs as well as D and clock inputs. The 4013 differs from the 7474, however, in that direct inputs become active when HIGH. If unused, therefore, these inputs should be tied to *ground*, instead of + 5 volts, as in the case of the TTL version. This is exactly the opposite protocol from the TTL.

In the clocked mode, the 4013 behaves much like the 7474. It is a positive-edge triggered device, so data is transferred to the Q output when the clock line goes HIGH. The action occurs on the positive-going *transition* of the clock pulse. In this respect, the 4013 and 7474 follow the same rules.

In the direct mode, the operation is similar to the 7474 but requires pulses of the opposite polarity. The rules for the set and clear inputs of the 4013 are as follows:

☐ If the set input is made HIGH, then the Q output goes HIGH and not-Q goes LOW.
☐ If the clear input is made HIGH, then the Q output goes LOW and not-Q goes HIGH.

J-K FLIP-FLOPS

The J-K flip-flop is very similar to the other clocked flip-flops, although certain differences exist that make the J-K special. Figure 7-19 shows a gate logic diagram for a J-K flip-flop, while the usual schematic symbol is shown in Fig. 7-20. Table 7-3 shows the truth table. Like the TTL and CMOS type-D flip-flops discussed earlier, the J-K flip-flop is capable of operating in either direct or clocked modes.

Direct operation uses the set and clear inputs to force the Q and not-Q outputs into specific states—HIGH or LOW. The truth table for TTL J-K flip-flops (7473, 7476, etc.) is shown in Table 7-4. The rules for these devices are:

☐ LOW conditions on both set and clear inputs results in a forbidden, or disallowed, state.
☐ Making the clear input HIGH and the set input LOW forces Q HIGH.
☐ Making the clear input LOW and the set input HIGH forces Q LOW.

Table 7-3. J-K Flip-Flop Truth Table.

INITIAL CONDITIONS				FINAL CONDITIONS	
OUT		IN			
Q	$\overline{Q}$	J	K	Q	$\overline{Q}$
0	1	0	0	0	1
0	1	0	1	0	1
0	1	1	0	1	0
0	1	1	1	1	0
1	0	0	0	1	0
1	0	0	1	0	1
1	0	1	0	1	0
1	0	1	1	0	1

Fig. 7-19. J-K flip-flop made from NAND gates.

Fig. 7-20. Circuit symbol for a J-K flip-flop.

□ Making both set and clear inputs HIGH causes the flip-flop to operate in the clocked mode.

Again the CMOS version operates in a similar manner, but with opposite polarity pulses. The CD4027 is a dual J-K flip-flop, so:

□ If both set and clear are LOW, then normal clocked operation is obtained.
□ Making clear LOW, and set HIGH, forces the Q output HIGH.
□ Making clear HIGH, and set LOW, forces Q LOW.
□ Making both set and clear HIGH is a disallowed condition.

Clocked operation means that all changes occur synchronously with the clock pulse. The J-K flip-flop operates on the negative going transition of the clock pulse, in distinct contrast to the behavior of the 7474 (type-D) device discussed earlier.

Table 7-5 shows the operation of TTL J-K flip-flops in the clocked mode. The changes that are shown occur as illustrated on

Table 7-4. Truth Table for Unclocked Operation.

SET	CLEAR	Q
0	0	(disallowed)
0	1	1
1	0	0
1	1	(normal condition for clocked operation)

Fig. 7-21. IC flip-flops.

105

92C5-17187R1

Table 7-5. Truth Table for Clocked Operation.

SET	CLEAR	CLOCK	Q
0 0 1 1	0 1 0 1	⌐▼⌐	No change 0 1 Goes to opposite

the negative transition of the clock pulse. The rules of operation are as follows:

- ☐ If both J and K are LOW, no change occurs in the output, regardless of clock pulse transitions.
- ☐ If J is LOW and K is HIGH, the Q output is forced LOW.
- ☐ If J is HIGH and K is LOW, Q is forced HIGH.
- ☐ If both J and K are HIGH, the Q output will go to the *opposite* of its present state; for example, if it is HIGH, then it will go LOW, and vice versa.

In the clocked mode, the CMOS 4027 device follows these same rules, except that the changes occur on the positive-going transition of the clock pulse.

J-K Flip-Flop Examples

Three examples of commercial IC J-K flip-flops have been mentioned: 7473 (TTL), 7476 (TTL), and 4027 (CMOS). Because the rules for these devices have already been given, they will not be repeated here. Figure 7-21 shows the pinouts for the 7473, 7476, and 4027 devices, respectively.

Chapter 8

Some Advice

Technicians servicing electronic equipment very often find themselves trying to figure out how the circuit works. Unfortunately, much equipment is on the market with little or no documentation. In one case, a supplier of a major medical computer system sold a complicated installation and then could not supply either service or *operator's* instruction manuals! Some manuals are little better than no manual at all. We have all seen those situations where we are at first elated to see a genuine service manual, only to find out a short time later that the service manual is almost useless. Some manufacturers seem to have a rather bad attitude towards servicers, especially independents. Their attitude is reflected by the nature of their manuals and other "field support." Note that most of these firms are not exactly hostile towards the field service industry, but simply don't care, unless their customers throw a fit over service problems. If you are a business person seeking to establish relationships with digital equipment manufacturers that will allow you to service their equipment, ask first to see the existing documentation. Find out whether or not you can live with their idea of "field support." This might well affect your contract negotiations. If the circuit level documentation is poor or scanty, for example, you might want to be responsible only for board swapping in the field, leaving all of the actual component level troubleshooting to factor or "depot" hands. But if they are able to provide a lot of detailed information about their product, in the form of schematics, PC board layouts and circuit operation descriptions, then consider offering them service to the component level.

POOR DOCUMENTATION

But right now, the principal concern is to try to figure out what to do in those cases when the documentation is too poor for usefulness, or is nonexistent, and the job *has* to be done regardless. That's not an easy order to fill, and nobody has any panacea.

The first advice is to ask you to learn as much as you can about the individual circuit elements; i.e., logic gates, flip-flops, etc. Buy all of the standard IC device reference manuals on digital chips. This publisher will be happy to send you their latest catalog of books. Also buy all of the manufacturer's data books on logic ICs. Motorola, Texas Instruments, National Semiconductor, Signetics, RCA, Intersil, Fairchild Semiconductor, and others publish comprehensive data books. A lot of duplication is among them, but they are all worthwhile in their own right. The reason is that *no one* makes completely universal lines of ICs. Each might have a few holes not covered by a data book. Also, each might have a few special "house number" devices not covered in other manufacturers' data books. Your library could save you an awful lot of trouble. A $200 investment in books every couple of years is not too great to ask of a business. After all, it will yield much in the way of time savings and completed jobs.

Another good piece of advice is to learn *well* the operation of circuits. For this, you might want to start with this basic textbook. Read as many books and magazine articles as time permits. Read them even if you are not particularly interested in the topic of the article itself. All circuits are constructed of building blocks. The device itself may be of no interest to you at the time, but the author might describe circuit techniques and building blocks that *are* of interest.

You might have guessed that I am a fan of reading and recommend it highly for everybody. I have noticed over the past two decades that the best service technicians are those who read hobby and trade magazines and books. This might seem a little bit self-serving, coming from an author, but it is true. I invite you to look around at the successful techs that you know.

In almost all cases, even in asynchronous circuits, the key to figuring out what is happening is *timing*. In all digital circuits, timing is critical. If the circuit is too complex for you to analyze in your head, then draw a timing diagram of how it should work. You have seen these diagrams all through this book. Use the rules of operation for the individual devices and make a diagram of the expected pulses and logic levels. Confirm them on an oscilloscope. This procedure does take time but is sometimes your only hope. Enshrine this concept deeply in your mind:

TIMING IS THE KEY!

Certain relationships exist between the different binary logic functions (OR, AND, NAND, NOR, XOR, NOT). These are described mathematically using logical notation, something called Boolean algebra and a tool called deMorgan's theorem. I do not believe that these matters are necessary in the training of good digital service technicians, but they are useful, if only for the purpose of helping you gain insight into how some circuits operate. We are not going to discuss the mathematical aspects here but do recommend that you consider reading a book that does cover such matters.

Do not be afraid to ask manufacturer's service personnel for advice on certain problems. They usually know the equipment better than anyone else. Some companies, though, seem to have a standing policy of referring all incoming telephone calls to the *sales department!* They are useless to you in most cases. Customer service offices are sales departments in disguise that have nothing at all to do with the kind of service department that you need. The exception is when you need to manipulate their built-in clout to gain access to a technician. Try to bypass sales and management (even the service manager) and get on the phone with someone who repairs the equipment—a real, live, hands-on person. Service department personnel are usually less inhibited about talking about their equipment problems than are sales and management people.

Incidentally, the engineering department is occasionally useful to field servicers, especially in solving really exotic problems. Most of the time, however, the engineering department is of little use to the servicer. It's not that they are stupid, of course, or too theoretical; they are too far removed from normal day-to-day service problems. There is also a built-in bias on the part of in-house engineering departments that everybody in the field is stupid. That's another problem you don't need. Besides, most engineers are too busy with *future* products to be much concerned over current or past products.

SOMETHING FOR TV SHOPS

The modern digitally controlled solid-state color TV receiver has caused some changes in the service industry. Some color TV sets now contain substantial amounts of previously unfamiliar digital electronics circuits, especially in tuning, remote control, and on-screen data display. The complexity of the sets has increased substantially over the years. This situation places increasing demands on the service personnel. No longer can an old-fashioned "tube jockey" repair substantial numbers of the shop's total work load. The same level of knowledge is now required for a TV servicer as was formerly required only in commercial, industrial, and military electronics work. Yet at the same time, TV set prices have not increased appreciably over the years. To the servicer, these factors mean that a more expensive technician is required to repair a less expensive product. This usually makes the customer less inclined to pay a reasonable price for service.

One answer to this dilemma is for the service shop to take its highly qualified people and go into some other branch of service work. The explosive growth of digital equipment and its seemingly ubiquitous distribution makes digital electronics a natural consideration. You will find the commercial, industrial, and institutional markets (i.e., hospitals, governments, etc.) more able and willing to pay for good quality service. The going rate for a TV service call in the Washington, DC area is around $30, and that is collected only with some trepidation. At the same time, commercial service rates are at a $40 to $55 per hour level. The difference becomes even more pronounced when you consider that the $30 collected by the TV servicer is a flat rate, one-time charge for the entire call. The commercial servicer operates on a portal-to-portal basis, and *he* gets to say how far away the portal was! This means that on an average two-hour service call the commercial tech collects at least $80, while the TV tech with about the same level of expertise collects only $30. There is little wonder why commercial work pays both the company and the technician more money.

Commercial electronics servicing has both advantages and disadvantages. One advantage is that your shop facilities may be located in lower rent warehouse sites or industrial park sites, rather than in high-priced suburban shopping malls. I know one company that operates out of the basement loading dock/garage area of a shopping center. The rent is less than *one-fifth* of the rent of the same amount of floor space upstairs in the mall. Not all are that good, however! You have to have a decent front to build a good, positive image in the mind of the consumer. Most commercial service contacts are by telephone or in person. There is practically no walk-in trade at all.

But that in-person bit means that your image problem is merely transferred from the physical plant to the technicians. In consumer service work, the techs can get away with open-collar sports shirts and dungarees. Commercial work, however, often demands a snappy, professional look. It has been claimed that magazine publisher and foreseer of much of today's electronics gadgetry, Hugo Gernsback, coined the term *servicer* to replace the *repairman* title used by radio technicians. Now there is ample justification to once again change titles—this time to *field engineer, service engineer,* or *electronic technologist,* which is the one I personally like. (In most states, the word *engineer* is reserved only for professional engineers licensed to practice engineering).

The importance of names, titles, and the personal appearance of the technicians is more than a little ego trip. It is critical to the way you are regarded by the client. They are less likely to pay higher prices for the services of someone who looks like an office cleaner. In this market, professionalism is required, and one must look professional. Consider, for example, some service departments of the big companies, such as IBM, Xerox, Digital Equipment Corporation, etc. I typed this section on an IBM *Selectric II* typewriter. In the year or so that I have owned it, IBM has serviced it three times. All of the persons ("persons" because one was a woman) who came to my house to repair, adjust or install my *Selectric II* have been dressed in *business suits.* And typewriter repair can be *dirty* work! When it came time to handle something messy, each would put on an apron; however, they always reported in proper business attire. They even carried their tools in a briefcase, rather than an ordinary tool box.

Chapter 9

Digital Counters:
Devices and Circuits

A digital counter is a device or circuit that operates as a frequency divider. The most basic counter is the simple J-K flip-flop connected such that J and K are tied HIGH. This arrangement makes the output produce *one* output pulse for every *two* input pulses. It is therefore a *binary*, or *divide-by-two*, *counter*. We can generalize, saying that a digital counter is a circuit that produces a single output pulse for x number of input pulses.

You are probably familiar with digital frequency counters, which are test instruments normally used to measure frequencies from radio transmitters. These instruments contain *decade*, or divide-by-10, counters. The generic term "counter," however, applies to a wide range of circuits that fit the definition in the previous paragraph.

There are two basic classes of digital counter circuits, *serial* and *parallel*. The serial types are all examples of *ripple* counters, which means an input change must ripple through all stages of the counter to its proper point. Parallel counters are also called *synchronous counters*. Ripple counters are operated serially, which means that the output of one stage becomes the input for the following stage.

The basic element used in counters is the J-K flip-flop, which is shown in Fig. 9-1. Note in the figure that the J and K inputs are tied HIGH and therefore remain active.

Fig. 9-1. J-K flip-flop wired for clocked operation.

A timing diagram in Fig. 9-2 shows the action of this circuit. When a J-K flip-flop is connected as in Fig. 9-1, its output changes state on the negative-going transition of the clock pulse. In Fig. 9-2, the first negative-going clock pulse causes the Q output to go HIGH. It will remain HIGH until the input sees another negative-going clock pulse, at which time the output will drop LOW again.

The actions required to complete the output pulse take two input clock pulses. This J-K flip-flop, therefore, divides the clock frequency by two, making it a *binary* counter.

A binary ripple-carry counter can be made by cascading two or more J-K flip-flops, as shown in Fig. 9-3. This particular circuit uses four J-K flip-flops in cascade. Any number, however, could be used. A problem with this simple type of counter is that only division ratios that are powers of two (2^n) can be accommodated. In all cases, the division ratio will be 2^n, where *n* is the number of flip-flops in cascade. In this circuit, therefore, there is a maximum division ratio of 2^4, or 16. Of course, if we can get to the Q outputs

Fig. 9-2. Operating waveforms. Note the F/2 output.

Fig. 9-3. Cascaded J-K flip-flops form a counter.

(ALL J&K INPUTS TIED HIGH)

of all four J-K flip-flops, we can create frequency division ratios of 2, 4, 8 or 16.

Frequency division is one use for such a counter. You may, for example, want to build a *prescaler* or some other type of divider. You can use a circuit such as shown in Fig. 9-3 for any division ratio that is a power of two (2, 4, 8, 16, 32 . . . 2^n).

But prescaling and most other simple frequency division jobs are but one example of counter applications. We can also use the circuit to store the total number of input pulses; the job most people mean when they think of "counters." Consider again the circuit of Fig. 9-3 and the timing diagram of Fig. 9-4. Outputs A, B, C, and D are coded in *binary*, with A being the least significant bit and D being the most significant bit. These outputs are weighted in the following manner: $A = 2^0$, $B = 2^1$, $C = 2^2$, $D = 2^3$. This is called the 1-2-4-8 system because these figures evaluate to $A =$

Fig. 9-4. Timing diagram of cascaded J-K flip-flops.

114

$1, B = 2, C = 4$, and $D = 8$. Recall from our discussion in Chapters 2 and 3 that these are the weights assigned to the binary number system. By arranging the digits in the form $DCBA$, a binary number is created that denotes the number of pulses that have the input.

Consider the timing diagram of Fig. 9-4. Note that all Q output changes occur following the arrival of each pulse. After pulse No. 1 has passed, the Q_A line is HIGH and all others are LOW. This means that the binary word on the output lines is 0001_2 (1_{10}); *one* pulse is passed.

Following pulse No. 2, we would expect 0010_2 (2_{10}) because *two* pulses have passed. Note that Q_B is now HIGH and that all others are LOW. The digital word is, indeed, 0010_2. If you follow each pulse, you will find the binary code to be as shown in Table 9-1.

The counter shown in Fig. 9-3 could be called a *modulo-16 counter, a base-16 counter,* or a *hexadecimal counter.* All of these terms mean substantially the same thing.

The output of a hexadecimal counter can be decoded to drive a display device that indicates the digits 0 through 9. A through F—the hexadecimal digits. In most applications where someone is to read the output, however, a *decimal* counter is needed. We human beings use the decimal number system because of our 10 fingers.

DECIMAL COUNTERS

A decimal counter operates in the base-10, or *decimal,* number

Table 9-1. Binary Code.

After Pulse No.	Word
0	0000
1	0001
2	0010
3	0011
4	0100
5	0101
6	0110
7	0111
8	1000
9	1001
10	1010
11	1011
12	1100
13	1101
14	1110
15	1111

system. The most significant bit of decimal counter produces one output pulse for every 10 input pulses. Decimal counters are also sometimes called *decade counters*. The decimal counter forms the basis for digital event, period and frequency counter instruments.

The hexadecimal counter shown in Fig. 9-3 is not suitable to decimal counting unless modified by adding a single 7400 NAND gate. Recall that a TTL J-K flip-flop uses inverted inputs for the *set* and *clear* functions. As long as the *clear* input remains HIGH, the flip-flop will function normally. When the *clear* input is momentarily brought LOW, the Q output of the flip-flop goes LOW.

The decade counter shown in Fig. 9-5 is connected so that all four *clear* inputs are tied together to form a common clear line. This line is connected to the output of a TTL NAND gate, which is one section of a 7400 IC device. Recall the rules of operation for the TTL NAND gate: If either input is LOW, then the output goes HIGH, but if both inputs are HIGH, then the output goes LOW.

The idea behind the circuit of Fig. 9-5 is to clear the counter to 0000 following the tenth input pulse. Let's examine the timing diagram of Fig. 9-6 to see if the circuit performs the correct action. Up until the tenth pulse, this diagram is the same as for the base-16, or hexadecimal, counter discussed previously.

The output of the NAND gate will keep the *clear* line HIGH for all counts through 10. The inputs of this gate are connected to the B and D lines. The D line stays LOW, which forces *clear* to stay HIGH, up until the eighth input pulse has passed. At that time (T_o in Fig. 9-6), D will go HIGH, but B drops LOW. We still have at least one input of the NAND gate (line B) LOW, so the *clear* line remains HIGH.

Fig. 9-5. Decade counter.

Fig. 9-6. Timing diagram a decade counter.

The *clear* line remains HIGH until the end of the tenth input pulse. At that point (T_2) both B and D are HIGH, so the NAND gate output goes LOW, clearing all four flip-flops. This forces them to go to the state where all Q outputs are LOW. The counter is now reset to 0000.

The reset counter produces a 0000 output, so both B and D are now LOW, and this forces the *clear* line HIGH once again. The entire reset cycle occurs in the period (T_3-T_2). This period has been expanded greatly in Fig. 9-6 for simplification. Actually, the reset cycle takes place in nanoseconds, or at the most, microseconds—whatever time is required for the slowest flip-flop to respond to the clear command.

The eleventh pulse will increment the counter one time, so the output will indicate 1 (0001). This counter, then, will count in the sequence 0-1-2-3-4-5-6-7-8-9-0-1-. . . . The output code is a 10-digit version of four-bit binary, or hexadecimal. It is called *binary coded decimal* (BCD).

SYNCHRONOUS COUNTERS

Ripple counters suffer from one major problem: lack of speed. The counter elements are wired in cascade, so an input pulse must ripple through the entire chain before it affects the output. A synchronous counter feeds the clock inputs of all flip-flops in parallel. This results in a much faster circuit.

Figure 9-7 shows the partial schematic for a synchronous binary counter. Synchronous operation is accomplished by using four flip-flops, with their clock inputs tied together, and a pair of AND gates.

One AND gate is connected so that both Q1 and Q2 are HIGH before FF3 is active. Similarly, Q2 and Q3 must be HIGH before FF4 is made active. On a clock pulse, any of the four flip-flops scheduled to change will do so *simultaneously*. Synchronous counters attain faster speeds, although ripple counters seem to predominate the common counter applications.

PRESET COUNTERS

A preset counter increments from a preset point other than 0000. For example, suppose we wanted to count from 5_{10} (0101_2). We could preset the counter to 0101_2 and increment from there. The counter output pattern will be:

5	0101	9	1001	3	0011
6	0110	0	0000	4	0100
7	0111	1	0001	5	0101
8	1000	2	0010	.	.

Figure 9-8 shows a common method for achieving preset conditions: the *jam input*. Only two stages are shown here for the

Fig. 9-7. Synchronous binary counter.

118

Fig. 9-8. Preset inputs using jam technique.

sake of simplicity, but adding two additional stages will make it a four-bit counter. Of course, any number of stages may be cascaded to form an N-bit preset counter.

In Fig. 9-8, the preset count is applied to *A* and *B,* and both bits will be entered simultaneously when clock line *CP2* is brought HIGH. Line *CP2* is sometimes called the *enter* or *jam* input. Once the preset bit pattern is entered, the counter will increment from these with transitions of clock line *CP1.*

DOWN AND UP-DOWN COUNTERS

A *down counter* decrements, instead of increments, the count for each excursion of the input pulse. If the reset condition is 0000, then the next count will be 0000 – 1, or (1111). It would have been 0001 in an ordinary up-counter. The count sequence for a four-bit down counter is:

0	0000	11	1011	6	0110	1	0001
15	1111	10	1010	5	0101	0	0000
14	1110	9	1001	4	0100	.	.
13	1101	8	1000	3	0011	.	.
12	1100	7	0111	2	0010	.	.

119

Basically the same circuit is used as before, but toggle each flip-flop from the not-Q of the preceding flip-flop. An example of a four-bit binary down counter is shown in Fig. 9-9. Note that the outputs are taken from the Q outputs of the flip-flops, but toggling is from the not-Q.

The *preset* inputs of the flip-flops are connected together to provide a means to preset the counter to its initial (1111) state. This counter is also called a *subtraction counter,* because each input pulse causes the output to decrement by one bit.

A *decade* version of this circuit is shown in Fig. 9-10. As in the case of the regular decade counter (the up-counter), a NAND gate is added to the circuit to reset the counter following the tenth count. The states are detected when C and D are HIGH, and the two middle flip-flops are cleared. This action forces the output to 1001_2 (9_{10}). The counter then decrements from 1001 in the sequence:

9	1001	5	0101	1	0001
8	1000	4	0100	0	0000
7	0111	3	0011	9	1001
6	0110	2	0010	.	.

UP/DOWN COUNTERS

Some counters will operate in both up and down modes, depending upon the logic level applied to a *mode* input. Figure 9-11 shows a representative circuit, in which the first two stages of a cascade counter are modified by the addition of several gates. If the mode input is HIGH, then the circuit is an UP counter. If the mode input is LOW, however, then the circuit is a DOWN counter.

Fig. 9-9. Four-bit down binary counter.

Fig. 9-10. Decade down counter.

TTL/CMOS EXAMPLES

Very few digital circuit designers construct counters from individual flip-flops. Too many ready-built IC counters are available in all of the major IC logic families.

There are three basic counters commonly available in the transistor-transistor logic (TTL) line: 7490, 7492 and 7493. The 7490 is a decade counter, the 7492 is a divide-by-12 (also called modulo-12) counter, and the 7493 is a base-16 (hexadecimal) binary counter. All three of these counters are of similar construction, and their respective pinouts are shown in Fig. 9-12. Other more sophisticated TTL counters are also available.

The 7490 is a *biquinary* type of decade counter. This means that it contains a single, independent, divide-by-two stage. This is followed by an independent divide-by-five stage. Decade division is accomplished by cascading the two stages.

Fig. 9-11. Up/down counters.

121

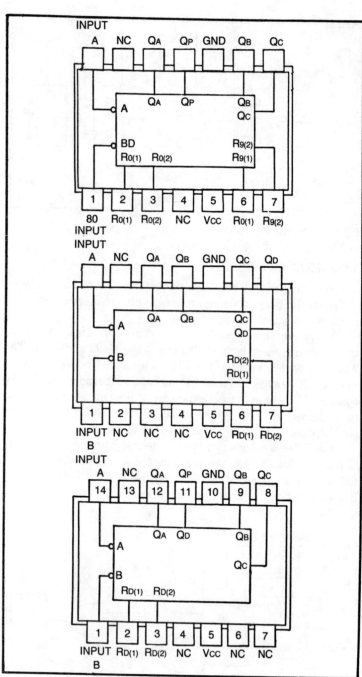

Fig. 9-12. TTL IC counters.

122

Both 7492 and 7493 follow similar layout schemes. In both, the first stage is a single divide-by-two flip-flop, followed by divide-by-six (7492) or divide-by-eight (7493) stages. These form divide-by-12 and divide-by-16 counters respectively.

The 74142 is a special function TTL IC that contains a divide-by-10 counter (BCD), a four-bit latch circuit, and a display decoder suitable for driving a *Nixie*® (registered trademark of the Burroughs Corporation). The 74142 is housed in a standard 16-pin DIP package.

Figure 9-13 shows the pinouts for the TTL type 74160 through 74163 devices. These are BCD and binary four-bit synchronous counters:

- □ 74160—Decade (BCD) synchronous, direct-clear.
- □ 74161—Binary, synchronous, direct-clear.
- □ 74162—Decade (BCD), fully synchronous.
- □ 74163—Binary, fully synchronous.

These counters typically operate to 32 MHz and dissipate approximately 325 mW. All are housed in standard 16-pin DIP packages.

Fig. 9-13. 74160 to 163 counter.

These four counters are different from those that have been discussed previously, because they are divide-by-N counters, where N is an integer. The value N is applied to the *data inputs* and loaded into the counter when the *load* terminal is momentarily brought LOW.

Examples of basic CMOS counters are shown in Fig. 9-14. Again, these examples are not exhaustive, but merely representative of those commonly used in electronic circuits. Not one of them is a multidigit decimal counter such as the eight-digit, 10-MHz device made by Intersil, Inc.

☐ 4017. This device is a fully synchronous decade counter, but the outputs are decoded 1-of-10. The active output is HIGH while the inactive outputs are LOW. The 4017 is positive-edge triggered. The *reset* and *enable* inputs are normally held LOW. If the *reset* is momentarily brought HIGH, the counter goes immediately to the zero state. The *enable* input is used to inhibit the count without resetting the device; that is, if the *enable* input is made HIGH, the count ceases and the output remains in its present state. The output terminal produces a pulse train of $F_{in}/10$, which is HIGH for counts 0, 1, 2, 3, and 4, and LOW for 5, 6, 7, 8, and 9.

☐ 4018. This device is a synchronous divide-by-N counter, where N is an integer of the set 2, 3, 4, 5, 6, 7, 8, 9, and 10. It is difficult to decode the outputs of this counter, so its principal use is in frequency division. For normal running, the *reset* and *load* inputs must be held LOW. The 4018 is positive-edge triggered. The N-code for determining the division ratio is set by connecting the input terminal to an appropriate *output* or in certain cases, an external AND gate. For *even* division ratios, no external gate is needed; merely connect the *input* terminal as shown in Table 9-2.

The odd division ratios, such as 3, 5, 7, and 9, require an external, two-input AND gate. The 4018 outputs are connected to the AND gate inputs, and the AND gate output is connected to the 4018 input. See Table 9-3.

The feedback line just described is also the main output from the counter. If, for example, "input" pin No. 1 will be (by the table above) 1/6 of the clock frequency.

The 4018 can also be parallel loaded using the jam terminals

Fig. 9-14. CMOS counters.

125

Table 9-2. Input Terminal Connections to the 4018 Counter.

N	Connect Input (Pin No. 1) To	Pin No.
2	Q1	5
4	Q2	4
6	Q3	6
8	Q4	11
10	Q5	13

P1 through P5. These terminals will program the 4018. A LOW on a jam input forces the related Q output HIGH, and vice versa. For example, if a LOW is applied to *P2,* it will force Q2 HIGH.

☐ 4022. This device is an octal, or divide-by-eight counter that provides 1-of-8 decoded outputs. The 4022 is very nearly the same as the 4017, which is a decade version (see the discussion for the 4017).

☐ 4026. The 4026 is a decade counter that produces uniquely decoded outputs for seven-segment displays. The 4026 is a positive-edge triggered device that is fully synchronous. This chip is similar to the 4017, in that it provides an F/10 output in addition to the seven-segment decoded outputs. The decoded outputs are HIGH for *active.*

There are two *enable* inputs. One is a *clock enable* input, which will cause the count to cease when brought HIGH. The counter outputs, however, remain in their present state when the clock is inhibited. The other enable input is a *display enable* terminal. A HIGH on this input will turn the display on, and a LOW will turn the display off. A possible use of this terminal is to turn off the high-current display when it is not needed.

☐ 4029. The 4029 is an up-down counter that will divide by

Table 9-3. Output Terminal Connections to the 4018 Counter.

N	Connect To AND Gate Inputs	Pin Nos.
3	Q1, Q2	5, 4
5	Q2, Q3	4, 6
7	Q3, Q4	6, 11
9	Q4, Q5	11, 13

either 10 or 16, depending upon whether pin No. 9 is HIGH or LOW. A HIGH on pin No. 9 causes the 4029 to be a base-16 binary counter. A LOW causes it to be a base-10 decade counter.

The count direction (up/down) is determined by the level applied to PIN No. 10. If pin No. 10 is HIGH, then the 4029 operates as an up-counter, but if pin No. 10 is LOW, then it operates as a down-counter.

Chapter 10

Display Devices and Decoders

People require certain types of displays to make the data presented by digital electronic circuits intelligible. Most people, it seems, are addicted to decimal readouts, and many find binary readouts almost unintelligible.

SIMPLE BINARY READOUT

A single incandescent lamp or a light-emitting diode (LED) can form a one-bit binary readout. In most cases, the LED/lamp will be driven through an inverter (see Fig. 10-1A), or a transistor (Fig. 10-2B).

In the case of Fig. 10-1A, a TTL open-collector inverter, such as the 7405 or 7406 devices, is used to drive an LED indicator. A series resistor between the LED anode and the V+ supply limits the current to a value that can be handled safely by the LED.

When the input signal is LOW, the output of the inverter is HIGH. As a result of this condition, the LED will be turned off (LOW = off). If the input is HIGH, the output of the inverter will be LOW, or (grounded). A current will flow through the LED, causing it to be lighted. In other words, HIGH = on.

In some cases, a single NPN transistor (2N3904, 2N3393, etc.) is used as the switch for the LED. The V+ supply is usually +5 VDC, but it can be any reasonable voltage. With appropriate transistor types and voltage sources, the circuit of Fig. 10-1B can

Fig. 10-1. Logic level detector using 7405/7406 inverter at A, level detector using transistor at B, and binary display.

also accommodate ordinary incandescent lamps, high-current incandescent lamps, and neon glow lamps (NE-2, NE-51, etc.).

The operation of the transistor driver is very straight forward:

☐ If the input is LOW (grounded), Q1 is cut off, so no current can flow through the LED (LOW = off).

☐ If the input is HIGH, then Q1 is turned on hard (it is saturated). This places the cathode end of the LED at near-ground potential (HIGH = on).

N-BIT BINARY READOUT

The simple binary output indicators of Fig. 10-1 can be extended to multiple-bit binary circuits. We need one driver per Figs. 10-1A and 10-1B for each bit of the binary number that we wish to create. In Fig. 10-1C, an eight-bit binary display has each LED representing a single, weighted, binary digit. The normal 1, 2, 4, 8, 16, 32, 64, 128 sequence is used. In this case, the 1, 4, 16, and 128 bits are lighted (HIGH), so the binary number being represented is 10010101_2, or $(128 + 16 + 4 + 1) = 149_{10}$. This type of display was used on primitive digital electronic instruments and is still used to display data in the registers of certain minicomputers and microcomputers.

SIMPLE DECADE DISPLAYS

Most electronic instruments use a *decimal* display instead of a binary display. This makes the data more readable.

129

Figure 10-2A shows a decade adaptation of the simple binary display. Recall that many decade counters use four J-K flip-flops and have outputs weighted 1-2-4-8. If a driver/LED combination is placed at each output bit, it will denote whether the bit is HIGH or LOW. By arranging all four bits of each decade in vertical columns, as in Fig. 10-2A, the count can be decoded. In the example, the MSD (1000s) reads 0011_2 (3_{10}), the 100s decade reads 0101_2 (5_{10}), the 10s decade is 0000_2 (0_{10}), while the units decade is 1001_2 (9_{10}). The count, then, is 3509_{10}.

An alternative column-type readout is shown in Fig. 10-2B. In this case 10 lamps represent the 10 decimal digits 0 through 9. In the two-digit case shown, the tens digit is 2 and the units is 7. The count, therefore, is 27.

The BCD display of Fig. 10-2A could be driven directly by the Q outputs of the counter flip-flops. Only a driver, such as shown in Fig. 10-1, is needed to interface the counter stage and the LEDs or lamps. The decade arrangement of Fig. 10-2B requires a *decoder* circuit, however. In this case, when interfacing a standard four-bit BCD counter, the decoder will be described as a BCD-to-one-of-10 decoder. This means that there are 10 outputs, and that one will become active for each of the 10 BCD digits. An equivalent circuit is shown in Fig. 10-3. It is in the form of a single-pole—ten-position (SP10T) rotary switch. The decoder in a real digital circuit, of course, would be an integrated circuit device. More on this follows. In effect, however, they do the same job as the rotary switch of Fig. 10-3.

If the BCD data at the four-bit inputs of the decoder is 0100_2 (4_{10}), then decoder output No. 4 will be grounded. This will cause LED No. 4 to be lighted. Other types of displays require different decoders. Only the Nixie® tube uses the 1-of-10 decoder.

NIXIE TUBES

The *Nixie* tube, made by the Burrough's Corporation, was among the first "people-engineered" digital displays. People are trained from earliest childhood to use the Arabic digits of our decimal system. We all find it difficult to decode columns of lights rapidly in our brains, but any display that produces an Arabic digit is almost instantly recognized.

The Burrough's Nixie tube produces an Arabic numeral corresponding to the BCD word applied to the input of a Nixie decoder circuit or IC decoder. The decoder is a high-voltage version of the 1-of-10 circuit.

Fig. 10-2. Decade binary-weighted display at A, and decade unweighted display at B.

131

Fig. 10-3. BCD-to-1-of-10 decoder analogy.

A Nixie tube is a neon glow lamp in which the cathodes are wire filaments formed into the various shapes of the 10 digits. A high voltage of 170 VDC is applied to a common anode terminal, so when any of the individual filaments are grounded, the neon gas particles close to their surfaces will glow orange, thus outlining the digit. The orange light emitted will take the shape of the digit. In the example shown in Fig. 10-4, the "4" cathode is grounded, so the digit lighted will be 4.

The Nixie tube suffers from two major defects: parallax and high voltage. The parallax problem exists because all 10 digits inside the tube are in slightly different planes. Although Fig. 10-4 shows the digits spread out, they are in reality placed one behind the other, so that they can be viewed from the front. This causes viewers who are positioned at an angle to see a distorted view. Typically, some digits will be seen fully, while others are distorted almost to the point of invisibility.

The Nixie tube has been all but eclipsed by more modern devices in newly designed equipment. Many instruments on the market still use Nixie tubes, however.

SEVEN-SEGMENT READOUTS

The seven-segment readout is probably the most common type

+170 VOLTS

DECODER

Fig. 10-4. Nixie® tube display.

of display used in digital electronics. It is used in calculators, frequency/period counters, and most other forms of instruments.

The basic seven-segment readout is shown in Fig. 10-5A and consists of seven lighted bars, or segments, labeled a through g. The decoder will cause appropriate segments to be lighted, forming the numeral. The light patterns for the 10 decimal digits are shown

Fig. 10-5. Seven-segment readout and seven-segment versions of the decade digits.

in Fig. 10-5B. Certain alphabet characters can also be represented, but only a few are easily recognized with the rest being quite poor representations of the actual letters. The 5 × 7 dot matrix of the next section will represent alphabetic characters more easily.

The seven-segment readout is available using several different illumination technologies. The RCA Numitron, for example, uses seven incandescent filaments to form the segments. These tubes are available in two sizes. The larger is housed in a nine-pin version of the small (usually 7-pin) vacuum tube envelope. The larger is in a standard 9-pin vacuum tube envelope.

The most common seven-segment display device uses LEDs to form the lighted segments. These are the red displays seen on so many electronics watches, calculators, etc.

The LED type does not usually offer the necessary brightness for viewing across the room or in a too-well lighted area (try viewing an LED display in sunlight). The gas discharge seven-segment display is brighter and can be larger, but only at the expense of requiring 150 to 200 VDC. This type of display produces a bright orange display color. On the other hand, the fluorescent display requires 20 to 40 VDC and produces a blue-green display color.

The display types mentioned above require substantial amounts of electrical current, up to 50 milliamperes per segment. If an "8" is illuminated, using all seven segments, it would draw as much as 8 × 50 mA, or 400 mA, from a DC power supply—all for one digit. An eight-digit digital frequency counter would then require 8 × 400 mA, or 3.2 A, to display the frequency 88888888.

The liquid crystal display, however, uses very little current. These displays are the dark gray-on-light gray types. Some use a yellow filter to provide a yellowish background. This improves the contrast for the viewer.

Liquid-crystal displays are *electrostatic* devices that only draw current when *changing state*. The static current drain is so small that CMOS calculators using liquid-crystal displays can operate as long as 1500 hours (one year of normal use) on a single pair of hearing aid batteries.

DECODERS

The binary or BCD code produced by a digital counter is not capable of directly driving the decimal display device. Some form of *decoder* circuit is needed to convert the four-bit binary, or BCD, to the one-of-ten or seven-segment codes used by the displays.

134

Fig. 10-6. 7442 decoder.

All of the decoders can be made to use assorted gates, properly interconnected, but it is far easier to use some of the various TTL and CMOS IC decoders that are available.

The 7442

The TTL device shown in Fig. 10-6 is a BCD-to-1-of-10 decoder. The 7442 outputs remain HIGH, except for the one selected by the BCD code applied to the inputs. The *selected* output goes LOW. The 7442 outputs can sink up to 16 mA of current at TTL voltage levels. A companion device, the 7445, will sink up to 80 mA and operates at voltages up to 30 VDC (output terminals only—the rest of the circuits inside the 7445 require the +5 VDC normally used in TTL).

The pinouts for the 7442/7445 are shown in Fig. 10-6, while the equivalent internal circuit is shown in Fig. 10-7. The internal operation is not very important to your understanding; it is included

V_{CC} = Pin 16
GND = Pin 8
O = Pin Numbers

Fig. 10-7. 7442 decoder schematic.

135

here merely as an example of how decoding can be accomplished. No other internal circuits will be shown, unless there is a compelling reason to do so.

The 7447, 7448, and 7449

These TTL devices operate as BCD-to-seven-segment decoders. The pinouts for the 7447 are shown in Fig. 10-8A, while those for the 7448 and 7449 are shown in Figs. 10-8B and 10-8C, respectively.

Like the previous example, an active output goes LOW when the appropriate BCD input word is applied. Unlike the 7442/45, however, a seven-segment decoder may have more than one output LOW at any given time. For example, if the BCD code 0110_2 were applied to the inputs, decimal "6" ($0100_2 = 6_{10}$) would be displayed. This means that the outputs controlling the c, d, e, f, and g segments of the seven-segment display would have to be LOW.

The 7447 can sink up to 40 mA of current at each output, and the outputs can handle up to $+30$ V. The package supply, however, must be $+5$ VDC.

The 7447 may be used directly on incandescent and fluorescent seven-segment displays, provided that the output current and voltage requirements are met. If LED seven-segment displays are used, however, a current limiting resistor must be connected in series with each output. If a V+ supply of $+5$ VDC is used, then 330-ohm resistors are used. As higher voltages are used, though, proportionally higher resistances must be used.

The test, or LT, terminal on the 7447 is used to provide a test of the decoder and the seven-segment display. If this pin is grounded, the display will read 8. If you examine Fig. 10-5B again, you will see that an 8 requires that all seven segments be lighted. This feature allows a quick visual test of the system in order to rapidly spot dead segments.

The RBI and RBO terminals are the *ripple blanking input* and *ripple blanking output*, respectively. These are used to turn off the display in order to eliminate leading zeroes in a multidigit display. If the blanking input is LOW, the display is turned off whenever the BCD code 0000_2 is applied to the inputs. The blanking output of the previous stage provides the LOW condition that turns off the display.

The 7447 is probably the most commonly used seven-segment decoder in the TTL line. The 7446 device is similar to the 7447, but uses open-collector outputs that require a pull-up resistor or

Fig. 10-8. 7447 decoder at A, 7448 decoder at B, and 7449 decoder at C.

segment load. The 7448 uses internal 2000-ohm pull-up resistors and has a pinout similar to the 7447. It will, however, sink less current and must operate at +5 VDC.

137

DISPLAY MULTIPLEXING

The display device in any digital instrument may require more current than the rest of the electronic devices in the circuit put together. One incandescent seven-segment display, for example, requires 40 mA per segment. Consider an eight-digit frequency counter in which the count is "88888888" (all seven segments on all eight readouts turned on). The current requirement is (8 × 7 × 40) mA = 2240 mA = 2.24 amperes! Clearly, then, digital readouts eat up a lot of current. If an instrument is to be battery-powered, then the display severely limits operating time.

One answer is to use *display multiplexing* to reduce the total *on-time* of each display device. This saves a large fraction of the energy that a steady-state display would consume.

Multiplexing is a technique whereby each digit is turned on in sequence, with all others being off. If the switching is done fast enough, such as at a 200-kHz rate, each display will be on only part of the time, yet because of the "Flicker fusion" phenomenon the display *seems* to remain on all of the time. The persistence of the viewer's eye—the latent image retained for a split second after the light actually extinguishes—makes the display appear steady.

Fig. 10-9. Electrical analogy for a multiplexed display.

138

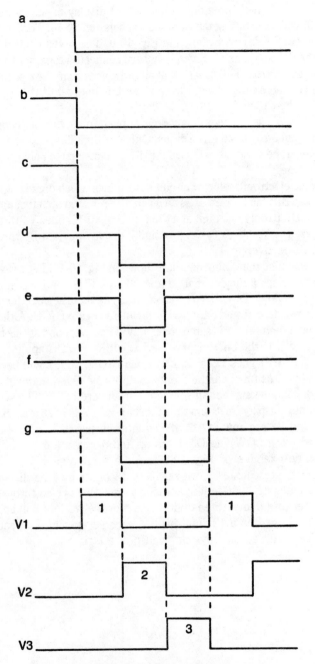

Fig. 10-10. Timing diagram of a multiplexed display.

139

A simplified three-digit multiplexed display is shown in Fig. 10-9, while the timing diagram is shown in Fig. 10-10. Three seven-segment displays are shown in Fig. 10-9. The seven control lines to the decoders, on all seven lines, are connected together to form a seven-line *bus*. This means that all three lines from the *a* segments are connected together to form the *a* bus line, all of the *b*'s, and all of the *c*'s, and so on.

When a seven-segment code is applied to the bus, it is applied simultaneously to all three digits. Only one digit lights up at a time, however, because the +5 VDC supply is connected to only one digit at a time.

An electronic selector circuit determines which digit is lighted. This circuit can be viewed as an electronic rotary switch that applies +5 VDC to only one digit at a time. In Fig. 10-10, lines V1 through V3 are the V+ lines (+5 V) to digital display devices D1 through D3, respectively.

Examine the timing diagram shown in Fig. 10-10. Let's say you want to display the decimal number "784". When +5 V is applied to V1, making V1 is HIGH, the counter outputs the code for 7. In other words, *a*, *b*, and *c* are LOW and all others are HIGH. Although all three digits see this same code, only V1 can respond at this time, because it is the only display device with +5 VDC applied.

When V1 goes LOW and V2 goes HIGH, digit 2 will become energized. At this time the counter outputs the seven-segment code for 8. This means that all segment outputs are LOW. The second digit now displays an 8 and the others are off. When V3 goes HIGH (V1 and V2 are now LOW), the counter output code is for 4 (*b*, *c*, *f*, and *g* are LOW), so D3 lights up and displays a 4.

The example of Fig. 10-9 is for a three-digit circuit. In a calculator or counter, however, 8 to 12 digits may be displayed simultaneously. These longer displays must be cycled through at a faster rate than a three-digit display, or the readout will appear to be dimmed; in a 12-digit display, each seven-segment readout device is turned on only one-twelfth of the time.

Chapter 11

Registers

A flip-flop is a circuit that is able to store a single *bit* of digital data. When two or more flip-flops are organized into some configuration to store multiple bits of data, they constitute a *register*. Most registers are merely specialized arrays of ordinary flip-flops.

There are several different circuit configurations that one would call registers. We classify these according to the manner in which data is input to and output from the register. We have, for example, *serial-in-serial-out* (SISO), *serial-in-parallel-out* (SIPO), *parallel-in-parallel-out* (PIPO), and *parallel-in-serial-out* (PISO) registers.

SISO AND SIPO

Figures 11-1 and 11-2 represent both SISO and SIPO shift registers. The only really significant difference is the parallel output lines used on the SIPO device; these lines would be absent on the SISO register. Note that some registers are both SISO and SIPO, depending upon how you use them. The SIPO shift register consists of a cascade chain of type-D flip-flops, with the clock lines tied together; i.e., they share a common clock line.

Recall one of the rules for Type-D flip-flops: Data can be transferred from the D-input to the Q-output *only* when the *clock* input is HIGH. This rule can be applied to the situation shown in Fig. 11-2, where the transmission of a single data bit from left to right through the SISO shift register is shown.

Fig. 11-1. SIPO and SISO shift register.

At the occurrence of the first clock pulse, the input is HIGH. This point is the D-input of FF1, so a HIGH is transferred to the Q1 output. This HIGH, which is also applied to the D input of the second flip-flop (FF2), remains after the clock pulse vanishes.

When the second clock pulse arrives, FF2 sees a HIGH on its D-input, and FF1 sees a LOW on its D-input. This situation causes a LOW at Q1 and a HIGH at Q2.

The third clock pulse sees a LOW on the D-inputs of both FF1 and FF2, and a HIGH at the D-input of FF3. The third clock pulse, then, causes Q1 and Q2 to be LOW and Q3 to be HIGH.

Fig. 11-2. Timing diagram of SIPO and SISO shift register.

142

Note that the SISO input remains LOW after the initial HIGH during clock pulse No. 1. This means that the single HIGH will be propagated through the entire SISO shift register, one stage at a time. The HIGH bit will shift one flip-flop to the right each time a clock pulse arrives. If the data at the input had changed, then the bit pattern at the input will be propagated through the shift register.

The shift register shown in Fig. 11-1 is a five-bit, or five-stage, shift register, although any bit-length could be selected. On the sixth clock pulse, therefore, the HIGH is propagated out of the register, so all flip-flops are LOW; i.e., $Q1 = Q2 = Q3 = Q4 = Q5 = \emptyset$.

The SISO shift register can be made into a SIPO device by adding parallel output lines, one each for Q1, Q2, Q3, Q4, and Q5. One use for the SIPO shift register is serial-to-parallel code conversion. For economic reasons, digital data is usually transmitted from device to device as a serial stream of bits, especially if long distances are involved. In other words, the bits of the digital word are sent over a communications channel one at a time. But most computers and other digital instruments use data in the more efficient parallel form. This format is both faster and more expensive; hence, the use of serial in transmission. As an example, consider an eight-bit binary data system. We would need an eight-stage SIPO to convert the eight-bit serial binary code to eight-bit parallel form. The code is entered into the SIPO register one bit at a time, so that after eight clock pulses, the first bit will appear at Q8, and the last bit at Q1.

PARALLEL

Parallel entry shift registers are faster to load than the serial input types because a given bit in the shift register can be changed directly, without applying it through all of the other stages. The new bit need not ripple through all of the preceding stages to reach its destination. There are two basic types of parallel entry: *parallel-direct* and *jam*.

In parallel-direct entry, or simply parallel entry, shown in the partial schematic of Fig. 11-3, the register must first be cleared by setting all bits to LOW and first bringing the *reset* line momentarily LOW. The data that is applied to inputs B1 through B_n can then be loaded into the register by momentarily bringing the *set* line HIGH.

The jam entry circuit of Fig. 11-4 is able to load data from the B1 through B_n inputs in a single operation. This job is accom-

Fig. 11-3. Parallel entry shift register.

plished by loading the complements of the B1 through B_n data into the other inputs. In actual IC shift registers using this technique, inverter stages are connected internally to the inputs to automatically complement the input lines. The outside user never sees the complementing process.

A recirculating shift register circuit is shown in Fig. 11-5. These registers are able to readout the data in serial format and then re-

Fig. 11-4. Jam entry shift register.

144

Fig. 11-5. Recirculating shift register.

enter it so that it is retained. This is done by connecting the output of the SISO register back to the input. As the data is read out of an ordinary SISO register, it is destroyed and lost forever; however, we sometimes wish to retain the data and be able to read it out, too. The solution here is to read it right back in as it is read out. Although only three bits are shown here for simplicity, the bit length could be almost anything.

One interesting application for the recirculating shift registers is in non-fade medical oscilloscopes. Most of the human physiological signals that physicians want displayed on a scope are very slow; i.e., 1 Hz represents a heart rate of 1 BPM. The trace tends to fade before the entire waveform is displayed. Modern oscilloscopes overcome this problem by digitizing the ECG waveform and then storing it in a recirculating shift register. By scanning the memory, which means reading out the serially stored data to a digital-to-analog converter, the entire waveform can be displayed on the CRT screen at one time. This is done 64 to 256 times per second in commercially available models. New waveform data is usually added to the shift register in four-bit groups.

IC EXAMPLES

Few designers would bother to use a series of cascade flip-flops in making a shift register. There are simply too many different types of IC shift register available on the market. Both TTL and CMOS examples are found.

Chapter 12

Timers and Multivibrators

A timer is a circuit that is able to produce an output level for a precise length of time. This output level may be normally HIGH and will drop LOW for the specified period of time; or, alternatively, it could be normally LOW, and then snap HIGH for the specified period when active. Devices of both sorts are known.

Many timers are also part of a family of circuits called multivibrators. In the case of the timer described above, we are talking about a *monostable multivibrator* or one-shot circuit. There are also *bistable* and *astable* multivibrators.

The names of these multivibrators are derived from the number of stable output states that each will allow. The monostable multivibrator, for example, has but one stable stage. It normally rests in its stable, or dormant, state. But when a trigger pulse is received, the output snaps to the active state. It will remain in the active state only for a given period of time, after which the output snaps back to the stable state.

The bistable multivibrator is the RS flip-flop. Recall from Chapter 7 that one trigger input on this stage will SET the output and make Q HIGH, while a trigger pulse on the other input will RESET the output or make Q LOW. The RS flip-flop, however, doesn't care which state it is in. It would remain happily in either SET or RESET condition; two stable states exist which is where the name *bistable* multivibrator comes from.

The astable multivibrator has *no* stable states. It will merrily flip and flop back and forth between SET and RESET conditions. The astable multivibrator, therefore, functions as a *clock generator,* producing a squarewave output wave train.

Both astable and monostable multivibrators function as timer circuits. Bistables, however, are not easily used in this capacity.

TIMERS VERSUS CLOCKS

It is usually the practice to distinguish *timers* from *clocks.* A timer circuit is the one-shot described earlier in this chapter. On the other hand, a clock is an astable multivibrator. Clocks are used to synchronize digital circuits.

There are several ways to generate a timer pulse. One of the most popular is the 555 integrated circuit timer. There is also a 556 device, which is a dual 555.

The 555 is inside an 8-pin miniDIP integrated circuit package. It has the advantage of being essentially free of period drift caused by variations in power-supply voltages; something that cannot be said of the unijunction transistor (UJT) relaxation oscillators previously used as timers.

The block diagram of the 555 is shown in Fig. 12-1, while the pinouts of the 8-pin miniDIP package is shown in Fig. 12-2. The 555 is one of the most popular IC devices used for timer circuits. Perhaps one of the reasons for this is that it is neither TTL nor CMOS, yet it can be interfaced with either of those logic families. By the way, the 555 is of bipolar transistor construction.

One principal difference between the 555 and TTL devices is

Fig. 12-1. Block diagram of 555 timer IC.

Fig. 12-2. Pinouts and package of 555.

that the 555 can be operated over a wide range of potentials; i.e., 5V to 15V, although potentials between 9V and 12V seems optimum. When the output terminal is LOW, the output terminal (pin No. 3) can *sink* up to 200 mA. When the output is HIGH, on the other hand, it will *source* up to 200 mA.

The 555 is capable of operating in either monostable or astable modes. The major difference between these modes of operation is the use of the output signal to retrigger the device.

The 555 is one of those nice devices that will allow the clever designer to use a lot of imagination. Before you can fully understand the wide range of uses this chip can offer, however, it is necessary to know the inner workings intimately. For this reason, we are going to explain the modes of operation using modified versions of Fig. 12-1 in the illustrations.

Figure 12-3 shows a monostable multivibrator using the 555 IC timer. Figure 12-3 is the block diagram, while Fig. 12-4 is the circuit as it appears in a schematic diagram.

The heart of the 555 timer is an RS flip-flop (FF) that is controlled by a pair of voltage comparators. An RS flip-flop, you will

Fig. 12-3. 555 in a monostable circuit.

148

Fig. 12-4. Same circuit as seen in circuit diagrams.

recall, is a bistable circuit; that is, it has two stable states. In its initial state the Q output is LOW, and the not-Q is HIGH. If a pulse is applied to the FF-SET (S) input the situation reverses itself, and the not-Q output becomes LOW, while the Q is HIGH.

A *comparator* is a device that is capable of comparing two voltage levels, and issuing an output that indicates whether they are equal or unequal. A comparator can be made by taking an operational amplifier or other high gain linear amplifier without any negative feedback. In the 555, two comparators are designed so that their outputs go HIGH when the two inputs are at the same potential.

Under initial conditions, at time T_o, the not-Q terminal of the RS flip-flop is HIGH, and this biases transistor Q1 hard on, placing pin No. 7 of the 555 at ground potential. This keeps capacitor C1 discharged. Also, amplifier A1 is an inverter, so the output terminal (pin No. 3) is initially in a LOW condition.

Resistors R_a, R_b, and R_c are inside the 555 IC and are of equal value, nominally 5000 ohms. These form a voltage divider that is used to control the voltage comparators. The inverting (–) input

149

of comparator No. 1 is biased to a potential of:

$$E_1 = \frac{(V_{cc})(R_b + R_c)}{R_a + R_b + R_c}$$

$$E_1 = \frac{2}{3} \times V_{cc}$$

This means that the output of comparator No. 1 will go HIGH when the control voltage to IC pin No. 5 is equal to $2/3\text{-}V_{cc}$. Similarly, the same voltage divider is used to bias comparator No. 2. The voltage applied to the noninverting input of the second comparator is given by:

$$E_2 = R_a + R_b + R_c \times \frac{R_c}{V_{cc}}$$

$$= \frac{1}{3} \times V_{cc}$$

When the voltage applied to the trigger input (IC pin No. 2) drops to $1/3\text{-}V_{cc}$, the output of the second comparator goes HIGH.

The control voltage is, in this case, the voltage across capacitor C2. This capacitor charges through resistor R_a and will reach $2/3\text{-}V_{cc}$ in less than 1 ms after power is applied. If a short, negative-going pulse is applied to the trigger input at time T_1 in Fig. 12-4, then the output of comparator No. 2 will drop LOW as soon as the trigger pulse voltage drops to a level equal to $1/3\text{-}V_{cc}$. This puts the flop-flop in the SET condition and causes the not-Q output to drop to the LOW state.

A drop to the LOW state by the not-Q output at time T_1 causes two things to occur simultaneously: One is to force the output of buffer amplifier A1 HIGH, and the other is to turn off transistor Q1. This allows capacitor C1 to begin charging through resistor R1. The voltage across C1 is applied to the noninverting input of comparator No. 1 through the threshold terminal (pin No. 6). When this voltage rises to $2/3\text{-}V_{cc}$, comparator No. 1 will toggle to its HIGH state and will RESET the RS flip-flop. This occurs at time T_2 and forces the not-Q output again to its HIGH state.

The output of amplifier A1 again goes LOW, and transistor Q1 is turned on again. When Q1 is on, we find that capacitor C1 is discharged. At this point the cycle is complete, and the 555 timer is again in its dormant state. The output terminal will remain LOW until another trigger pulse is received. The approximate length of time that the output terminal remains in the HIGH condition is given by:

$$T = T_2 - T_1$$
$$T = 1.1 \ R1C1$$

where T is the time duration in seconds, R1 is the resistance of *R1* in ohms, and C1 is the capacitance of *C1* in farads.

This function is graphed in Fig. 12-5 for times between 0.01 and 10.0 seconds with values of R1 and C1 that are easily obtainable. The time relationship between the trigger pulse, the output pulse, and the voltage across capacitor C1 is shown in Fig. 12-6.

If the reset terminal is not used, it should be tied to V_{cc} to prevent noise pulses from jamming the flip-flop. If, however, negative-going pulses are applied simultaneously to the trigger input (pin No. 2) and the reset terminal (pin No. 4), the output pulse will terminate. When this occurs, the output terminal drops back to the LOW state, even though the time duration (T) has not yet expired.

An astable multivibrator is similar in many respects to the monostable variety, except that it is self-retriggering. The astable multivibrator has no stable states, so its output will swing back and forth between the HIGH and LOW states. This action produces

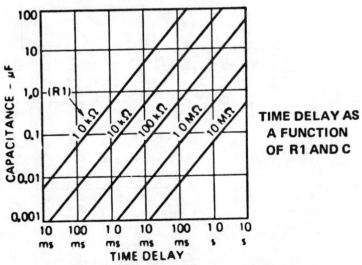

Fig. 12-5. Operating parameters of the 555 in monostable operation.

Fig. 12-6. Waveforms of the 555.

a wave train of squarewave pulses. An example of a 555 astable

$$T = T_1 + T_2$$
$$T = 0.693 (R1 + 2R2) C1$$

multivibrator circuit is shown in Figs. 12-7 through 12-9. Again, we have a version of Fig. 12-1 for block diagram analysis in Fig. 12-7, and the circuit as it will appear in schematic diagrams is shown

Fig. 12-7. 555 in astable operation.

152

Fig. 12-8. Same circuit as seen in circuit diagrams.

in Fig. 12-8. As in the previous case, the inverting input of comparator No. 1 is biased to $2/3\text{-}V_{cc}$, and the noninverting input of comparator No. 2 is biased to a level of $1/3\text{-}V_{cc}$ through the action of resistor voltage divider network R_a, R_b, and R_c. The remaining two comparator inputs are strapped together and are held at a voltage determined by the time constant $C1(R1 + R2)$. Under initial conditions, the not-Q output of the RS flip-flop is HIGH. This

Fig. 12-9. Timing diagram of 555 in astable operation.

153

turns on transistor Q1, keeping the junction of resistors R1 and R2 at ground potential. Capacitor C1 has been charged, but when Q1 turns on, C1 will discharge through resistor R2. When the voltage across capacitor C1 drops to a level of 2/3-V_{cc}, the output of comparator No. 1 goes HIGH and that resets the flip-flop. This action again turns on Q1 and allows C1 to discharge to 1/3-V_{cc}. Capacitor C1, then, alternatively charges to 2/3-V_{cc} and then discharges to 1/3-V_{cc}. Figure 12-9 shows the relationship between HIGH and LOW times. The high time, T_1, is given by:

$$T_1 = 0.693 \ (R1 + R2) \ C1$$

and the low time by:

$$T_2 = 0.693 \ (R2) \ C1$$

The total period of the waveform, T, is the sum of T_1 and T_2, and is given by:

$$T = .693 \ (R1 + 2R2) \ C1$$

In any electrical circuit, or any other physical system for that matter, the frequency of an oscillation is the reciprocal of the period. In this case, then, the frequency of oscillation is:

$$F_{Hz} = 1/T_{sec}$$

$$F_{Hz} = \frac{1}{0.693 \ (R1 + 2R2) \ C1}$$

$$F_{Hz} = \frac{1.44}{(R1 + 2R2) \ C1}$$

The last equation is shown graphically in Fig. 12-10 for frequencies between 0.1 Hz and 100,000 Hz, using easily obtainable component values. The relationship between the C1 voltage and the output state is shown in Fig. 12-11. The *duty cycle*, also called *duty factor*, is the percentage of the total period that the output is HIGH. This is given by:

$$DF = \frac{T_1}{T_1 + T_2}$$

Fig. 12-10. Parameters of the 555 in astable operation.

$$DF = \frac{R2}{R1 + R2}$$

The trigger input of the 555 timer is held in the HIGH condition in normal dormant operation. To trigger the chip and initiate the output pulse pin No. 2 must be brought down to a level of 1/3-V_{cc}. Figure 12-12 shows how this can be done manually. Resistor R4 is a pull-up resistor used to keep pin No. 2 HIGH. Capacitor C3 is charged through resistors R3 and R4. When S1 is pressed, the junction of R3 and C3 is shorted to ground, rapidly discharging C3. The sudden decay of the charge on C3 generates a negative-going

Fig. 12-11. Operating waveforms of the 555 in astable operation.

155

Fig. 12-12. Manual triggering of the 555.

pulse at pin No. 2 that triggers comparator No. 2, initiating the output pulse sequence.

The 555 timer is also available in a dual version called the 556. This IC is a 14-pin DIP package containing two timers, independent of each other except for the power-supply voltage. It is, then, a dual 555. Both 555 and 556 timers are so useful that it is an even bet you will someday use one or both if you do any circuit design work at all.

Another useful IC timer—in fact, a *family* of timers—is the Exar type XR-2240, XR-2250, and XR-2260. The internal circuitry of these is shown in Fig. 12-13. These devices are second sourced in the form of the Intersil 8240, 8250, and 8260, respectively. The 2240 is a binary counter with 8-bit outputs. The 2250 is a similar device, except that the outputs are in BCD code. The 2260 is also BCD, except that the most significant bit is 40, instead of 80. This allows the 2260 to be used as a timer with 60 count rather than 100. In the 2250 and 2260 devices, the regulator output terminal (pin No. 15) is used as an overflow that allows cascading of several stages. Table 12-1 shows the pinouts and their respective weighting.

Fig. 12-13. Pinout diagram of the XR-2240 timer.

These timers will operate over a supply voltage range of +4.5 to +18 VDC. The time base section is a clock circuit that is very similar to the 555 device. This similarity is shown in Fig. 12-14, which shows the internal circuitry of the XR-2240. One main difference between the 555 timer and the time base portion of the XR-2240 lies in the relative reference levels created by the respective internal voltage dividers (R1, R2, and R3 in Fig. 12-14). In the 555 timer all three resistors were of equal value, so comparator input terminals were at $0.33V_{cc}$ and $0.66V_{cc}$. In the Exar chip, on the other hand, the reference levels are $0.27V_{cc}$ and $0.73V_{cc}$, respectively. One result of this is simplification of the duty factor equation that gives the period of the waveform.

Table 12-1. Pinouts of the Exar Timers.

Pin No.	2240/8240	2250/8250	2260/8260
1	1	1	1
2	2	2	2
3	4	4	4
4	8	8	8
5	16	10	10
6	32	20	20
7	64	40	40
8	128	80	(N.C.)

157

Fig. 12-14. Block diagram of the XR-2240 timer.

The binary counter section consists of a chain of JK flip-flops connected in the standard manner, where each stage functions as a divide-by-2, or binary, counter. The binary counter chain is connected to the output of the time base section through an NPN open-collector transistor. The transistor collector is also connected to IC pin No. 14 (called *time base output*) so that a 20K ohm pull-up resistor can be connected between the collector and the output of internal regulator pin No. 15.

Digital outputs from this counter are, in the usual fashion given as voltage levels at a set of IC pins. Each output bit is delivered to a specific terminal of the IC package where it is connected to a pull-up resistor similar to that used for the time base output terminal. The output terminals will generate a LOW condition when active. This may seem to be opposite to the usually accepted arrangement, but there is a method to this madness, and it does create a highly versatile, stable, long-duration counter. These are properties not usually associated with single-IC designs.

Figure 12-15 shows the basic operating circuit for the XR-2240 timer IC. This chip proves interesting because the sole difference between circuits for astable and monostable operation is the 51K ohm feedback resistor linking the reset terminal (pin No. 10) to the wired-OR outputs. The timer is set into operation by application of a positive-going trigger pulse to pin No. 11. This pulse is routed to the control logic and has several jobs to perform simultaneously: resetting the binary counter flip-flops, driving all outputs low, and enabling the time base circuit. As was true in the 555 IC, this timer works by charging capacitor C1 through resistor R1 from a positive voltage source, $V+$ or V_{cc}. The period of the output waveform is given by:

$$T = R \times C$$

where T is the period in seconds, R is the resistance of R1 in ohms, and C is the capacitance of C1 in farads.

The pulses generated in the time base section are counted by the binary counter section, and the output stages change states to reflect the current count. This process will continue until a positive-going pulse is applied to the reset terminal.

Figure 12-16 shows the relationship between the trigger pulse, time base pulses, and various output states. The reason for the open-collector output circuit is to allow the user to wire a permanent OR output so that the actual output duration can be programmed. Each binary output is wired in the usual power-of-two sequence: 1, 2, 4, 8, 16, 32, 64, and 128. If these are wired together, the output will remain LOW as long as *any one output* is LOW. This allows the output duration to be programmed from 1T to 255T, where

Fig. 12-15. XR-2240 circuit.

159

Fig. 12-16. Timing diagram of the XR-2240 timer.

T is defined as in the last equation, by connecting together those outputs which sum to the desired time period. For example, design a timer with a 57-second time delay. In the binary notation, decimal 57 is equal to:

$$32 + 16 + 8 + 1 = 57$$

Set $T = R1C1 = 1$ second wire and wire together the pins on the XR-2240 corresponding to these weights, then a 57-second time delay will be realized. The base diagram to this IC shows us that these pins are numbers 1, 4, 5, and 6. If those four pins are shorted together, the counter output will remain LOW for 57 seconds following each trigger pulse.

The time base or the wired-OR terminals could be changed to vary the output periods. Of course, if the time base frequency were doubled, the counter would reach the desired state in half the time. This feature allows programming of the XR-2240 to time durations that might prove difficult to achieve using conventional circuitry.

Each output must be wired to V_{cc} through a pull-up resistor of 10K ohms, unless, of course, the wired-OR output configuration is used. In that case, a single 10K ohm resistor is used.

Current through the output terminals must be kept at a level of 5 mA or less. This serves as a general guide to the selection of pull-up resistors.

160

The amplitude of the reset and trigger pulses must be at least two PN junction voltage drops ($2 \times 0.7 = 1.4$V). In most practical applications, it might be wise to use pulses greater than 4V amplitude, or standard TTL levels, in order to guard against the possibility that any particular chip may be a little difficult to trigger near minimum values or that outside factors tend to conspire to actually reduce pulse amplitude at the critical moment.

Synchronization to an external time base or modulation of the pulse width is possible by manipulating pin No. 12. In normal practice, this pin, which is the noninverting input of comparator No. 1, is bypassed to ground through a 0.01 μF capacitor so that noise signals will not interfere with operation. A voltage applied to pin No. 12 will vary the pulse width of the signal generated by the time base. This voltage should be between +2V and +5V for a time base change multiplier of approximately 0.4 to 2.25, respectively.

If you wish to synchronize the internal time base to an external reference, connect a series RC network consisting of a 0.1 μF capacitor and a 5.1K ohm resistor to IC pin No. 12. This forms an input network for sync pulses, and these should have an amplitude of at least 3V at periods between 0.3T and 0.8T (see Fig. 12-17). Another way to link the count rate to an external reference is to use an external time base. This signal may be applied to the *time base output* terminal at pin No. 14.

Each Exar XR-2240 has its own internal voltage regulator circuit to hold the DC potentials applied to the binary counters at a level compatible with TTL logic. This consists of a series pass transistor, which has its base held to a constant voltage by a zener diode. If operation below 4.5 VDC is anticipated, it becomes necessary to strap the regulator output terminal at pin No. 15 to V_{cc} (V+) at pin No. 16. The regulator terminal can be used to source up to 10 mA to external circuitry or an additional XR-2240.

Fig. 12-17. Sync input.

LONG-DURATION TIMERS

Long-duration timers can be built using any of several approaches. You could, for example, connect a unijunction transistor (UJT) in a relaxation oscillator configuration, or use a 555. In almost all cases, though, there seems to be an almost inevitable error created by temperature coefficient and inherent tolerance limits of the high-value resistors and capacitors required. Also, a certain amount of voltage drop is across the capacitor due to its own internal leakage resistances and the impedance of the circuit in which it is connected. The use of the XR-2240 IC timer will all but eliminate such problems because a higher clock frequency is allowed. Easier-to-find component values will therefore satisfy the equation, $T = R \times C$. It is usually easier to specify and obtain high-quality, precision components in these lower values. This makes the period initially more accurate and results in less drift with changes in temperature.

An example of a long-duration timer is shown in Fig. 12-18; it consists of two XR-2240 IC timers cascaded to increase the time duration. In this circuit, the time base output of IC2 (pin No. 14) is an input for an external time base. Timer IC1—also an XR-2240—is used as a time base for IC2. The most significant bit of IC1 (pin 8, weighted 128) is connected to pin No. 14 of IC2. This pin will remain LOW from the time IC1 is triggered until time T_o = 128R1C1. It will then go HIGH and trigger IC2.

The binary counters in IC2 will increment once for every 128T. Timer IC1 is essentially operating in the astable mode because its reset pin is tied to the reset of IC2, and that point does not go HIGH

Fig. 12-18. Cascading to make a long-duration timer.

until the programmed count for IC2 forces its output to go HIGH.

The total time duration for this circuit under the conditions shown (IC2 input from pin No. 8 of IC1, and with all IC2 outputs connected into the wired-OR configuration) is 256^2, or 65,536T. You can, however, manipulate three factors to custom program this circuit to your own needs: time base period (R1C1), the output pin on IC1 used to trigger IC2, and the strapping configuration on IC2. In other words:

$$T_o = R1C1 \times T_o' \times T_o''$$

Where T_o is total period in seconds that the output is LOW, T_o' is the period that the selected output (s) of IC1 is LOW, and T_o'' is the period that the selected output (s) of IC2 is LOW.

For example, assume that the product R1C1 is one second, (as if R1 = 1 megohm and C1 = 1 μF). If the output remains LOW for a period of 65536T = 65536 (R1C1) = T_o, it will remain LOW for 65536 seconds, or about 18 hours. This is on a one-second pulse! Of course, it is impossible to accurately generate a time delay this long using any of the other techniques. It is, for example, relatively common to find high-value electrolytic capacitors necessary in long-duration RC timers that are rated with a -20 percent to $+100$ percent tolerance in capacitance. This is incompatible with the goal of making a precision time base of long duration. In addition, most electrolytics—and that includes tantalum—will change value over time while in service. For most filtering and bypassing applications, this is acceptable; in timing, though, it is deadly. The use of cascaded XR-2240 devices allows us to select more easily managed values of resistance and capacitance.

PROGRAMMABLE TIMERS

By adding some external circuitry, the 2240, 2250, and 2260 timers can be made programmable; i.e., an operator can select the time duration at will. Figure 12-19 shows a circuit that allows timer control with a digital word (binary for the 2240 and BCD for the 2250 and 2260). In this circuit, the output terminals are connected to a pair of 7485 TTL magnitude comparators.

The 7485 device is composed of a set of TTL exclusive-OR gates connected to compare two four-bit words and issue an output that indicates that word A is greater than B, A equals B, or that A is less than B. In this case, the output of the 2240/2250/2260 can

Fig. 12-19. Long-duration timer with preset count.

be compared with the word applied by a microcomputer output port or an external binary or BCD thumbwheel switch.

The timing diagram of Fig. 12-19 is shown in Fig. 12-20. Because word A programs the timer, the output durations are given in terms of the value of A, and the $R1C1$ time base period.

When a trigger pulse is received, word A has previously been set to some value between 00000000_2 and 11111111_2. All this time, a word A is either greater than word B or equal to it if word A is also 00000000_2. In the former case, the A-greater-than-B output from IC3 is HIGH and the other two outputs are LOW. When the counter has incremented so that word A and B are equal, then the A-equals-B output goes HIGH for a period of one clock pulse,

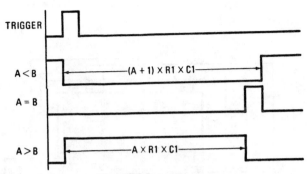

Fig. 12-20. Timing diagram of a long-duration timer.

after which word B is greater than word A, so the A-less-than-B output goes HIGH. The timing durations are $(A + 1)$ R1C1 for A-less-than-B, and AR1C1 for A-greater-than-B. The $A = B$ output produces a single pulse at time AR1C1.

Assume that the circuit shown in Fig. 12-19 is programmed so that word A is 178_{10} (10110010_2y) and that the RC time constant is five seconds. What is the duration of the A-greater-than-B pulse? The duration is A × R1 × C1, or (178_{10}) × (5_{10} seconds), or 890 seconds (about 14.8 minutes).

Thumbwheel switches can be used with binary, BCD, or octal output encoding to program these IC timers. Figure 12-21 shows an XR-2250 timer connected to a BCD thumbwheel switch. The BCD output lines of the switch are connected to the timer output lines, and the switches are, in turn, connected to pull-up resistor R1. The switches provide a simple way to build circuits such as Fig. 12-15 using convenient front panel switches to change the output duration as desired.

Fig. 12-21. Presettable one-shot.

V = PIN 16 ALL ICS

Fig. 12-22. Presettable counter in units of time (hours, sec).

166

Figure 12-22 shows the circuit of a precision 100-hour timer using the Exar XR-2250 and XR-2260 (or Intersil 8250/8260) timer ICs. Connected in cascade, four timers are programmed by thumbwheel switches.

Timer IC1 is used as a time base to generate 1-Hz clock pulses from the 60-Hz AC power line. The values of C1, C2, R1, and R2 depend upon the amplitude of the 60-Hz line used for the time base.

The XR-2260 counts to 59. On the 60th count, it generates a carry pulse from pin No. 15. This output pulse represents a frequency division of 60 for the XR-2260 and 100 for the XR-2250.

As shown in Fig. 12-22, the *second* and *minutes* sections of the circuit are thumbwheel-programmed, cascaded XR-2260 ICs, while the *hours* section is a thumbwheel-programmed XR-2250 IC, allowing a maximum of 99 hours. Both XR-2260 ICs use the carry output pulse to drive the time base input of the following stage.

The thumbwheel switches program the total duration of the output pulse, but some applications might require a continuous monitoring of the time that has *elapsed* since the trigger pulse started the counters. To do this, you can use an events counter or a clock IC that uses a 1-Hz input. Another alternative is to use an hours-minutes-seconds counter circuit with its reset terminal connected to the timer output. The timer output is active-low, so it will allow counting during its time duration but will reset the counter when it goes HIGH at the end of the time duration.

Chapter 13

Data Multiplexers and Selectors

The fastest way to transfer data within digital systems or between two different digital systems is to connect all of the like-bits in parallel—all B0 lines in parallel, all B1 lines in parallel, etc. But parallel data transfers are expensive to implement. They become terribly expensive in hardware costs when there is more than one data source. If there are, say, three channels of eight bits each, 24 wires or radio communications channels are necessary to transfer the data. While this may not be too much of a problem inside the equipment cabinet, it becomes one when renting cross-city or intercity telephone lines—the standard lines used for data communications. One seemingly standard 12-bit computer is fitted with eight I/O ports. To connect them in parallel another computer or other devices, 96 wires would be needed. Try paying for 96 cross-country telephone lines.

It might not seem like much. After all, any technician worth a lot load of salt can easily and correctly connect 96 wires in a very short order. But these wires must be viewed in a little different light once outside the same room. Digital signals cannot be passed along a piece of wire longer than a few feet unless certain conditions are met. The problem is the sharp rise time of the pulses used in digital electronics. Low-impedance line drivers and line receivers must be used at the input and output ends of the line, respectively. These "wires" must be viewed as *communications channels* instead of mere wires. When you go outside your own building and have to rely on the services of the telephone common carriers, then they are indeed communications channels. How would you like to pay for 96 radio or wire communications channels? The cost is simply too much for this type of data communications.

The solution is in *serial data transmission*, the subject of the next chapter. In this chapter, though, we will cover how to make several data channels fit into one communications channel, or how to make parallel data bits into serial data bits. The answer is in *multiplexing*.

MULTIPLEXING

Multiplexing is the art of *interleaving* data so that they can be transmitted together over a single line or through a single data communications channel. Two basic forms of multiplexing exist: *frequency domain* and *time domain*. The most common examples of frequency domain are color television, FM stereo broadcasting, and subcarrier telephony. All of these use one or more subcarriers to transmit information. Digital systems, however, greatly use time domain multiplexing.

Figure 13-1 shows a simplified example of a time domain

Fig. 13-1. Electrical circuit model of time domain multiplexers.

169

multiplexer circuit. Here two rotary switches select from any of six different data channels. These switches are rotated *synchronously* so that both are always in the same position; i.e., S1 is connected to channel No. 3 at the same time that S2 is connected to channel No. 3. This situation allows us to transmit data from six channels serially along the communications path by using techniques that will be discussed more fully in the next section.

Figure 13-2 shows a system in which four eight-bit devices are connected to a single eight-bit input port. If they were connected in parallel, 4×8, or 32, wires would be necessary. Because the system is being multiplexed, however, only one I/O port (instead of four I/O ports) and 10 wires are needed. Eight of the wires are used to form the eight-bit data bus over which all of the data is to pass. The remaining two wires are used for control of the multiplexer. In the circuit shown, the multiplexer can be viewed as a programmable switch. These two wires do the programming, according to the rules shown in Table 13-1. If the code is 00, then

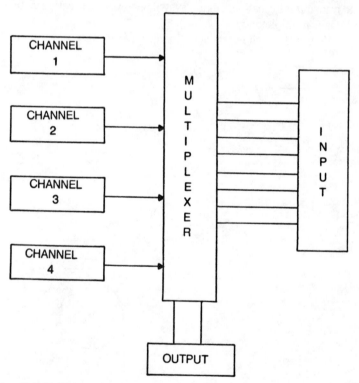

Fig. 13-2. Multiplexed system of four channels.

Table 13-1. Truth Table of a Multiplexed System.

Channel	Code
0	00
1	01
2	10
3	11

channel 0 is selected and connected to the data bus. If the code is 01, then channel 1 is selected, and so on. A simple two-bit binary counter, or a counter program if a computer is used, will create the codes. It will sequence 00-01-10-11-000 . . . until turned off or disabled.

The circuit shown in Fig. 13-2 assumes that the data at the inputs will always be ready, because it operates in a quasi-asynchronous mode. Synchronous operation would require at least one more wire. Of course, it must also be assumed that the device receiving the data will not change the data selector code until it has successfully completed the current data transfer. It will not, for example, switch to channel No. 2 until it has successfully input the data on channel No. 1.

Figure 13-3 shows another form of digital multiplexing, used extensively on calculators and other portable digital instruments. LED and incandescent seven-segment display readouts (see Chapter 10) draw a considerable amount of current from the power supply. Some devices draw up to 40 mA per segment. In the four-digit display shown in Fig. 13-3, when all segments are turned on (the displayed value is 8888), the current drawn is:

$$\frac{0.04 \text{ amperes}}{\text{segment}} \times \frac{8 \text{ segments}}{\text{digit}} \times 4 \text{ digits} = I$$

$$0.04 \times 8 \times 4 \text{ amperes} = 1.28 \text{ amperes}$$

If this were a 12-digit portable calculator, even more current capability (almost 4 amperes) would be needed. Once you begin to require so much current on a continuous basis, battery life drops rapidly. If a battery large enough to allow lengthy operation is used, then portability becomes questionable, at the least.

The answer to the battery life dilemma is to multiplex the digits. Only one digit at a time will be turned on, but if the switching rate

Fig. 13-3. Multiplexed seven-segment display.

is fast enough, the *persistence* of the human eye will make the display *appear* constant.

Figure 13-3 shows a multiplexed digital display, while Fig. 13-4 shows the timing diagram. Recall from Chapter 10 that a seven-segment readout requires one wire for each segment; these are usually labeled a,b,c,d,e,f, and g. In a typical device, power is applied to the anode of the digit, and the decoder will ground the terminals to those segments that are to be lighted. To display a 3, for example, ground terminals a,b,c,d, and g (see the inset of Fig. 13-3). The instrument will place the proper seven-segment code for the digit to be displayed on the seven-bit data bus at the appropriate time. Timing is critical to the proper operation of this system, and it is handled by the control logic section.

The *digit selector* is a circuit that enables each digit in sequence. In the crude example shown here, a digit is selected by applying +5 VDC to the common anode. This circuit has four output lines, one for each digit. When a selector line is HIGH, then +5V is applied to its digit.

In the situation shown in Fig. 13-3, the four-digit number 4137 is displayed. The sequential operation is as follows:

- ☐ During time T1, the seven-segment code that represents "4" is placed on the data bus. Simultaneously, the digit selector applies +5 VDC to digit No. 1. This causes digit No. 1 to display a "4."
- ☐ During time T2, the seven-segment code representing "1" is placed on the data bus. Simultaneously, the digit selector applies +5 VDC to digit No. 2 after having disconnected all other digits.
- ☐ During time T3, the seven-segment code representing "3" is placed on the data bus. Simultaneously, the digit selector applies +5 VDC to digit No. 3, which turns off the other three digits.
- ☐ During time T4, the seven-segment code representing "7" is placed on the data bus, and the digit selector turns on digit No. 4.

This sequence happens continuously. Only one digit is turned

Fig. 13-4. Timing diagram of the multiplexed seven-segment display.

on at any one time, but they all turn on sequentially many times per second.

A simple example of a two-channel multiplexer is shown in Fig. 13-5. This circuit uses a common CMOS digital integrated circuit, the 4016, or its newer and preferred cousin, the 4066. Both 4016 and 4066 devices are easily available through mail order electronics houses specializing in IC sales. Note that the CMOS electronic switches used in these devices are *bilateral*, meaning that current will pass in either direction. This means that they can be used as either analog, where bipolar signals might be encountered, or digital multiplexers.

Each of the switches inside the 4016/4066 devices is completely independent of the others. They share only common power-supply terminals.

The control pins are 5, 6, 12, and 13. When the voltage applied to these pins is equal to the voltage on pin No. 7, the switch selected by that particular pin is turned off. In digital systems, pin No. 7 will most likely be grounded, so the switches are turned off when their control pins are grounded (zero volts). The voltage applied to pin No. 7 can be anything in the range 0V to −5V (−7V in some models), but the negative voltage are only seen occasionally in digital circuits.

Applying a voltage equal to the voltage on pin No. 14 to a control pin will turn *on* the switch. In most digital circuits, this will be +5V, although CMOS devices can operate up to +18 VDC. The rationale is to make the whole circuit TTL compatible.

The terms "on" and "off" refer to the series impedance of the switch. In the off condition, the series impedance is very high, usually well over a megohm. In the on condition, however, the series impedance of the switches drops very low. It will be in all cases less than 2000 ohms, and in some devices less than 100 ohms.

In the "muxer" of Fig. 13-5, only two of the switches are used. This circuit, incidentally, is also useful for making a single trace oscilloscope into a dual trace scope. Note that one terminal of both S1 and S2 are connected together to form a common output. The remaining terminal on both switches goes to the respective channels. Pin No. 1 is connected to channel No. 1, and pin No. 4 is connected to the channel No. 2.

The switching action is performed by a J-K flip-flop connected in the clocked binary divider method. The Q and not-Q outputs on a J-K flip-flop are complementary. This means that one output will be HIGH when the other is LOW. When the state of the FF

174

Fig. 13-5. Two-channel multiplexer based on 4016/4066 device.

changes, the HIGH output becomes LOW and the LOW output becomes HIGH. The timing diagram shown in Fig. 13-5 shows this relationship more clearly.

During interval T1, the Q output is HIGH and not-Q is LOW. This turns on switch S1 and turns off S2 (pin No. 5 is HIGH and pin No. 13 is LOW). According to the rules for operation of the 4016/4066, channel No. 1 is then connected to the output.

Following the next clock pulse (J-K flip-flops operate on the negative-going edge of the clock pulse), the condition of the outputs becomes reversed. S1 is now turned off and S2 is turned on. This will connect channel No. 2 to the output.

The switches are turned on and off alternately. This means that the output will be connected to the respective inputs alternatively.

In the case of an oscilloscope switch, only this one circuit is needed to make things happen. In data communications systems, however, the signals are now interleaved, and that makes things a hopeless nightmare at the other end of the transmission path unless some form of decoding is provided. A *demultiplexer* must then be provided at the receive end of the path.

Happily, the 4016/4066 devices are bilateral. Basically the same circuit is useful for "demuxing". The serial data stream coming into the receiver is fed to the common terminal between the two switches (used as an output in the mux end and an input in the demux end).

Three additional standard CMOS devices are useful as

multiplexers and demultiplexers: 4051, 4052, and 4053. All three of these CMOS devices have chip enable pins that allow the chip to be turned on and off at will.

The 4051 shown in Fig. 13-6 is a 1-of-8 selector and can be viewed as analogous to a single-pole, eight-position rotary switch. Again, like the 4016/4066 devices, the switches are bilateral. Three control terminals (A/B/C) select which one of the switches is turned on at any given time. These are programmed according to the ordinary binary sequence for the digits 0 through 7, such as 000 on A/B/C causes channel 0 to be on, 001 causes channel 1 to turn on, and so on. A binary counter, driven by a clock, will sequence through all eight possible conditions.

The 4052 device shown in Fig. 13-6 is a little different. It contains two independent 1-of-4 switches. The independence is not total, however. There is only one set of control terminals: A/B. These respond to the codes for 0 through 3. When A/B sees 000,

Fig. 13-6. 4051, 4052 and 4053 analog/digital muxers.

PIN CONNECTIONS
AND ORDERING INFORMATION

A_2	A_1	A_0	E_N	"ON" CHANNEL
X	X	X	L	NONE
L	L	L	H	1
L	L	H	H	2
L	H	L	H	3
L	H	H	H	4
H	L	L	H	5
H	L	H	H	6
H	H	L	H	7
H	H	H	H	8

TOP VIEW

16 PIN HERMETIC DUAL IN-LINE
(Q Suffix)

MODEL	TEMP RANGE
MUX-88AQ	$-55°C$ TO $+125°C$
MUX-88BQ	$-55°C$ TO $+125°C$
MUX 88EQ	$-25°C$ TO $+85°C$
MUX 88FQ	$-25°C$ TO $+85°C$

Fig. 13-7. Precision Monolithics MUX-88.

both channel 0s are turned on. When the code is 01, both channel
1s are turned on. This process continues. We can view the 4052
as a pair of single-pole, four-position rotary switches.

The 4053 device shown in Fig. 13-6 is a triple 1-of-2 switch.
It is, therefore, analogous to a three-pole, double-throw switch, or,
perhaps, three SPDT switches ganged together.

All three of these CMOS devices can be used as multiplexers
and demultiplexers. In most case, though, the 4051 would be used
for a single channel. If you were to multiplex together more than
one data source, the other devices could be used.

IC MULTIPLEXERS/DEMULTIPLEXERS

The previous examples have been regular 4000-series CMOS
devices pressed into service as mux/demux devices. The trend in
recent years had been to use specialized integrated circuits that
employ FET technology for the muxer/demuxer. In this section,
certain commercial examples that have become popular are
covered.

Figure 13-7 shows the Precision Monolithics, Inc. (PMI)
MUX-88 multiplexer/demultiplexer device. It is a monolithic eight-
channel switch using *junction* field-effect transistors, instead of the
MOSFETs used in CMOS devices. This substantially reduces the
susceptibility of the device to damage by static electricity
discharges.

The MUX-88 is versatile enough to operate from either TTL or CMOS logic levels and can therefore be used in almost any existing application. The switching action is make-before-break.

As in the simple example of Fig. 13-5, the output of the MUX-88 is connected to one side of all eight switches. The remaining terminal of each switch will go to the respective channels. Like the 4051, this device responds to a three-bit (octal) binary code that determines which switch is to be turned on at any given time. There is also a chip enable terminal that turns on and off the MUX-88.

The Datel MX-series of multiplexer/demultiplexers is shown in Fig. 13-8. This family of IC devices is useful in a variety of applications because it is compatible with TTL, CMOS, and even the old DTL logic system. Like the MUX-88, these are monolithic integrated circuits, rather than discreet or hybrid technology

Fig. 13-8. Pinouts for MX-series muxers.

8	4	2	1	INHIB.	ON CHANNEL
X	X	X	X	0	NONE
0	0	0	0	1	1
0	0	0	1	1	2
0	0	1	0	1	3
0	0	1	1	1	4
0	1	0	0	1	5
0	1	0	1	1	6
0	1	1	0	1	7
0	1	1	1	1	8
1	0	0	0	1	9
1	0	0	1	1	10
1	0	1	0	1	11
1	0	1	1	1	12
1	1	0	0	1	13
1	1	0	1	1	14
1	1	1	0	1	15
1	1	1	1	1	16

Table 13-2. MX-1606 Channel Addressing.

devices. There are different sizes of multiplexers in this family, but all use either 2-, 3-, or 4-bit binary codes as the switch select signals. Tables 13-2 through 13-4 show channel addressing.

One advantage the specialized CMOS devices has over standard devices is that transfer accuracy is generally guaranteed more closely. In the Datal MX-series, for example, 0.01 percent transfer accuracy is obtainable at sample rates up to 200 kHz. The devices will tolerate bipolar signal excursions over the range ±10V, so they can be used for either analog or digital muxers.

This series requires a power supply in the range ±5 to ±20 VDC. Interestingly, the power consumption is only 7.5 mW

4	2	1	INHIB.	ON CHANNEL
X	X	X	0	NONE
0	0	0	1	1
0	0	1	1	2
0	1	0	1	3
0	1	1	1	4
1	0	0	1	5
1	0	1	1	6
1	1	0	1	7
1	1	1	1	8

Table 13-3. MX-808 and MX-807 Channel Addressing.

179

2 1	INHIB.	ON CHANNEL
X X	0	NONE
0 0	1	1
0 1	1	2
1 0	1	3
1 1	1	4

Table 13-4. MXD-409 Channel Addressing.

(standby condition). Even at a high sampling rate, in the 100-kHz range, the power consumption is only 15 mW.

Be careful about the total power *consumption* and the total package dissipation. The former refers to the amount of power required from the DC supply in order to power the internal circuits of the device. The total dissipation refers to the amount of power that can be dissipated by the device and includes the power loss in the switches. Because each switch has a series resistance—even though small—there will be a signal loss, and this translates as a power dissipation. In the Datel MX-series, the total allowable dissipation is on the order of 725 mW for the MX-808 and 1200 mW (that's 1.2 watts!) for the MX-1606 and MXD-807 devices.

The MX-1606 is a 1-of-16 channel single-ended multiplexer. You might view it as an electronic SP8T switch. It uses a four-bit address select code; four bits are required for 16 channels because $2^4 = 16$. Similarly, the four-channel device (MXD-409) and the eight-channel device (MXD-807) use two-bit and three-bit select codes, respectively.

The transfer accuracy of any electronic switching multiplexer depends upon the source and load resistances. The output voltage is given by the standard voltage divider equation:

$$E_{out} = \frac{E_{in} R_L}{R_s + R_{on} + R_L}$$

where E_{out} is the switch output voltage, measured across load resistor R_L; E_{in} is the input voltage, or the open-circuit voltage of the input signal source; R_L is the load resistance; R_s is the output impedance of the signal source; and R_{on} is the resistance of the switch when *on*.

It doesn't take a mathematical genius to figure out that it is wise to keep the R_s and R_{on} terms of the equation as low as possible, and to keep the R_L term as high as possible. We can meet the low-R_{on} criterion by using an operational amplifier with a low output impedance at the input of the electronic switch. The load resistance requirement can also be met by using an operational amplifier. It should have an input impedance that is greater than 10^7 ohms. This is not too hard to obtain because some low-cost BiFET or BiMOS input op amps boast input impedances in the 10^{12} ohms range.

☐ **Example.**

Find the output voltage error in percent if the following terms apply: $R_s = 100$ ohms, $R_L = 10^7$ ohms, and $R_{on} = 10^3$ ohms.

$$\frac{E_{out}}{E_{in}} = \frac{R_L}{R_s + R_{on} + R_L} = \frac{(10^7)}{(10^2) + (10^3) + (10^7)}$$

$$= 9.9989 \times 10^{-1} = 0.99989$$

Converting to percent:

$$S = \frac{1 - 0.99989}{1} \times 100\%$$

Fig. 13-9. Datel MX-series muxers.

$$= (0.00011) (100\%)$$
$$= 0.011\%$$

In this example, you can see that an error of approximately 0.01 percent will be obtained when we use values that are reasonably close to those actually obtained in the real world. To reduce this error even further, then, use an operational amplifier with a very high input impedance. If an RCA CA3130, CA3140, or CA3160 were used with a 10^{12}-ohm input impedance, the error would reduce to a ridiculously small value—so small that it is difficult to calculate on a 10-digit electronic calculator. Devices such as the LF156 are also used for this purpose.

Figure 13-9 shows a scheme for using four 16-channel muxes, such as the Datel MX-series (MX-1606), to make a 64-channel multiplexer. A 6-bit select code is required to determine the channel that is turned on at any one time. To minimize loading, the common outputs are connected together at the noninverting input of a unity gain operational amplifier.

Chapter 14

Data Transmission

Data transmission involves moving digital data from one location to another. The transmission path might consist of only a few yards across the room or down the hall. It might also consist of a long-distance communications system involving telephone (wire) communications, radio or microwave links, long-distance coaxial cable hookups, or even satellite communications. In this chapter, we are going to discuss some of the fundamental aspects of data communications, both local (inside the same building) and long distance.

PARALLEL DATA COMMUNICATIONS

Short-distance data communications can be carried out over parallel lines. The only requirement here is to have one wire for each bit and control signal needed to complete the job.

One of the simplest examples is the ordinary 7-bit ASCII keyboard interface to a computer. Such a keyboard interface receives the binary word for an ASCII alphanumeric or machine control character on the lower seven bits (B0 through B6) of the port. The high-order bit (B7) of the port is connected to the keyboard *strobe* output. The strobe becomes active when the data on the lower seven bits correctly represents the desired character. Such a circuit is shown in Fig. 14-1. The programmer of the computer will have to write a keyboard program that looks something like Fig. 14-2. The computer constantly tests for B7 being active (let's assume

Fig. 14-1. Connecting a keyboard to an input port.

active-HIGH). If B7 = LOW, then the program jumps back to the beginning and starts over again; it continues to loop through, testing for B7 = HIGH. But if it finds B7 = HIGH, the active condition in this example, the keyboard data is valid—only one strobe pulse is generated for each keystroke, so the program jumps to an input instruction. After the main program "does its thing" with the input data, the program jumps back to the beginning of the loop.

Figure 14-3 shows another possible situation involving parallel data transmission. In this case, two computers are connected together; it could just as easily be a peripheral and a computer—

Fig. 14-2. Program flow chart.

Fig. 14-3. Connecting two computers together.

the principle is the same. Such a circuit may or may not use the control signals, depending upon whether the data transfer is to synchronous or nonsynchronous. This circuit merely requires a bit of cable, usually flat ribbon cable in computer systems, and a couple of connectors. The programmer will write a subroutine program to interface the data transfer.

In the case of nonsynchronous, or asynchronous, data transfer, a loop program such as given for the keyboard is used. In fact, the keyboard represents an example of an *asynchronous peripheral*. Sometimes, though, the computer will be doing something else when the peripheral needs servicing. Alternatively, it could be that the two devices have vastly different operating speeds. This frequently occurs in the case of electromechanical peripherals sending or receiving data from computers. The data transfer might have to be synchronized in these cases. And the signals between the two devices need to be controlled to cause synchronization. This is sometimes called *handshaking*. By sending a pulse to device B, device A will tell device B, "Hey! I'm ready to receive the next data byte." By returning another pulse, Device B will reply, "OK, here it comes." While it is usually a little more complicated than this, the idea is the same.

185

SERIAL DATA CIRCUITS

Serial data bits are transmitted one at a time. Perhaps the simplest example of a serial data system is the old-fashioned Morse code. The operator would tap out the dots and dashes of the code one at a time, and that is how they would be transmitted along the system or over the air. Figure 14-4 shows an example of a serial digital data stream. In this case, the binary number 11001001 is being transmitted. The binary digits can take on only one of two values (HIGH or LOW) at any one time. The key is in knowing exactly when a given bit is expected and then looking to see the condition at that instant. Because of this requirement, it is critical to know when the word begins and when it ends. It is also critical that the speed of transmission at the receive end be matched within a very small error to the speed of transmission at the sending end. The beginning of the word will be denoted by the existence of a start bit. Ordinarily, the line will be held in one condition or the other; it then switches to the condition of the start bit when the transmission begins. The start of a word is very often indicated by a start bit that is 1.5 or 2 times the normal duration of the data bits.

Data speed is measured in bits per second, sometimes called *baud*. The precise definition of baud is somewhat more involved and subject to some differences of opinion, but bits per second is used most frequently. You will sometimes see the term baud rate, but because baud *is* a rate, it is redundant; nevertheless its use is so frequent that we must nod to standard practice and forget proper word usage.

It is necessary that the baud rate be matched at both ends of the serial data communications system. Because of this, it has become standard to use one of only a few different popular rates: 110, 300, 600, 1200, 2400, 4800, and 9600 baud. The slowest rate, 110 baud, is one of the more popular used in older printers and

Fig. 14-4. Serial representation of binary word 11001001.

teletypewriters. At 110 baud, it is possible to transmit approximately 10 characters per second, each character requiring 100 ms. At this rate, teletypewriter speeds of 100 words per minute can be accommodated, with time left to spare for such housekeeping chores as carriage return and line feed. As a general rule, most devices use baud rates of 1200 or less because it is possible to use ordinary telephone circuit. Faster rates require special conditioned lines that are a lot more costly. These special lines are often dedicated to only that one data communications link, thereby increasing the cost even further if you do not need the link all of the time.

Serial data communications systems can be classified into the following: simplex, half-duplex, and full-duplex. The *simplex* system permits communications in only one direction. This is like ordinary radio broadcasting or radio control applications, where a fixed transmitter at one end of the path communicates with a fixed receiver at the other end. Half-duplex is a system in which two-way communications is possible, but not simultaneously; only one direction at a time is used. This is analogous to two-way radio systems. Operator A can talk to B, but B cannot talk back until A is finished and relinquishes the air. Full-duplex is a system in which two-way communications can exist simultaneously. In those systems, the data can flow in both directions at once. Each end of the communications system will have its own receiver and transmitter that are completely independent of each other.

INTERFACE STANDARDS

Several different serial data interface standards have been adopted by the data communications industry. Among these are RS-232, RS-422, GPIB (general-purpose interface bus), and IEEE-488. The purpose of these standards is to provide standard voltage levels, operating speeds, and connector pinouts so that equipment made by different manufacturers will be directly compatible with each other. A computer manufacturer, for example, might provide an RS-232 interface card with the machine. This means that a peripheral manufacturer can sell equipment that will interface directly with the computer, simply making the peripheral to the RS-232 standard. A complete range of modems (modulator-demodulator), CRT video terminals, printers, teletypewriters, and other peripheral devices are available in RS-232. This is probably the most popular standard of the Electronic Industries Association (EIA).

Unfortunately, the RS-232 is one of the older standards used by the industry. It predates TTL, so the logic levels of the RS-232 system are not TTL compatible. This is something of a disadvantage, but common usage apparently keeps RS-232 alive and well.

The RS-232 standard specifies the logic level voltages and the impedances so that devices of different manufacturers have some small hope of operating together with success. Figure 14-5 shows the RS-232 voltage levels. Note that there are actually two versions of the RS-232 standard. The older standard, RS-232B, has a wider spread between the two voltage levels used to represent the two logic states. Logical-0 was present when the voltage was between +5V and +25V, while logical-1 existed when the line voltage was between -5V and -25V.

On the other hand, RS-232C uses narrower ranges. The logical-0 level was +5V to +15V, while logical-1 was -5V to -15V. This change resulted in speedier operation of the circuit: Under worst case conditions, it takes less time to make the transition from -15V to +15V than it does from -25V to +25V.

Designers of RS-232C circuits must contend with ±25V logic levels, driver output impedance specs, receiver input impedance specs, and a 30 Vμs slew-rate requirement. Unfortunately, TTL devices do not meet the logic level specs of RS-232C, yet they make up the bulk of the digital devices connected to RS-232C circuits. In most cases, designers use one of the ready-made IC level

Fig. 14-5. RS-232B and C voltage levels.

Table 14-1. RS-232 Pin Definitions.

Pin No.	RS232 Name	Function
1	AA	Chassis ground
2	BA	Data from terminal
3	BB	Data received from modem
4	CA	Request to send
5	CB	Clear to send
6	CC	Data set ready
7	AB	Signal ground
8	CF	Carrier detection
9	undef	
10	undef	
11	undef	
12	undef	
13	undef	
14	undef	
15	DB	Transmitted bit clock, internal
16	undef	
17	DD	Received bit clock
18	undef	
19	undef	
20	CD	Data terminal ready
21	undef	
22	CE	Ring indicator
23	undef	
24	DA	Transmitted bit clock, external
25	undef	

translators designed especially for RS-232C. The Motorola MC1488 line driver and MC1489 line receiver are examples.

The connector used for RS-232C is a standard type-D connector, commonly used for many different computer applications. This is a 25-pin connector that is readily available. It is sometimes called a DB-25 connector. At other times it is known as an RS232 connector. The latter is a misnomer that is commonly used in the digital industry. Table 14-1 shows the pinouts for the RS-232 25-pin standard connector.

TELETYPEWRITER CIRCUITS

Teletypewriters are electromechanical typewriters that will type under automatic control. Basically, such machines use a mechanism similar to an electric typewriter. Instead, though, a teletypewriter has a series of five or seven solenoids to pull in the selector bars of the typewriter for a desired character. The original teletypewriters were made by the Teletype Corporation of Skokie, IL. Note that *Teletype* is a trademark of the Teletype Corporation.

These machines operate using a series current loop for the electromagnets that pull in the selector bars. Figure 14-6 shows a recommended interface circuit for older 60 mA machines that allows computer control of the current loop. One bit (usually LSB) of a computer output port is used to control the circuit. It is applied to the base of switching transistor Q1. The user must provide a 130 VDC power supply and a rheostat (R2). With the circuit connected Q1 turned on, and a key on the teletype pressed to close the loop, R2 is adjusted for a current of approximately 60 mA.

Bit control is gained by turning Q1 on and off. When Q1 is turned on, the loop is closed, and current will flow. This is the logical-1 condition and is called a *mark* in teletypewriter terminology. Alternatively, when the input bit to Q1 is LOW, Q1 is turned off. The loop is now open, and no current will flow. This is the logical-0 condition, called a *space*. Diode D1 is used as a transient suppressor. The electromagnets of the teletypewriter will produce a sharp spike when the loop is opened due to inductive kick.

The simple circuit of Fig. 14-6 is too simple, and there are some unfortunate problems with it. One is the fact that the TTL output port of the computer is not entirely safe if Q1 develops a collector-to-base leakage current of any significant magnitude. The 130 VDC supply has been known to damage the output port under that condition. The second problem is that spikes from the circuit, those not entirely suppressed by D1, will get into the computer because the circuit is not well isolated. The modification of the circuit shown in Fig. 14-7 solves both of these problems. In this case, the tele-

Fig. 14-6. 60 mA teletypewriter current loop connection.

Fig. 14-7. Isolated version of a current loop connection.

typewriter loop control transistor is turned on and off by the LSB of the computer output port, but through an optoisolator. Such a device consists of a light-emitting diode (LED) and a phototransistor inside a common package. When the diode is dark, the transistor is not biased and will therefore not conduct any C-E current. When the LED is turned on, however, the base of the transistor is forward biased, so the transistor is turned on. A HIGH on the LSB of the computer output port will turn on the LED, causing the optotransistor to turn on. This will short out the bias to the loop control transistor (Q1), making it see a LOW. Of course, this circuit requires us to use inverted code from the computer, but that is no problem. We could also add one additional common emitter stage or even a TTL inverter to rectify the problem.

The circuit in Fig. 14-6 works only with the Baudot encoded, 60 wpm teletypewriters. These machines (Model 15, 21, etc.) are old and no longer currently sold. Of course, older systems and amateur radio operators using RTTY and amateur computerists will still use such machines. It is noted, however, that modern computers equipped with RTTY interface units are rapidly pushing teletypewriters off the air.

Most modern systems will use a more recent machine. One of the standards of the industry is the Teletype Model 33. This machine uses a 20 mA (instead of 60 mA) current loop. It is encoded in ASCII, which is more compatible with computer practices. The

Fig. 14-8. Connection to the control terminal block of a model 33 Teletype® .

model 33 is capable of 100 wpm (110 baud) operation. A circuit for interfacing the model 33 is shown in Fig. 14-8. The terminal strip shown is found on the rear of the teletypewriter, underneath the removable top cover, on the right-hand side as viewed from the front of the machine. It is necessary that you identify the pin numbers, especially plus pins 1 and 2 are carrying 110 VAC. These can be recognized as having white and black wires, respectively. This circuit does not use the internal 20-mA power supply; instead, it uses external power supplies. For proper isolation, however, it is necessary to use ±5V and −12V power sources separate from the computer power supplies. For even greater isolation, use separate power supplies *and* an optoisolator (Fig. 14-9). In Fig. 14-9 the transistor in the previous circuit is replaced with the transistor to add optoisolator protection to the *send* side of the circuit.

Later models of Teletype, such as the Model 43, are made more

Fig. 14-9. Modification of the model 33 for better isolation.

192

computer compatible. These are available in options that permit either 20-mA current loop operation or RS-232, ASCII operation.

UNIVERSAL SYNCHRONOUS
RECEIVER-TRANSMITTER CHIPS

The design of digital circuits for serial data transmission was once something of a chore. It required a complex selection of gates, registers, and other logic elements. Today, it's different. Some semiconductor manufacturers offer single LSI chips that will do the entire job called universal asynchronous receiver-transmitter (UART) chips.

Asynchronous transmission is preferred over synchronous transmission because it is not necessary to precisely track the clocks at each end. The clocks must be operating at very nearly the same frequency, but they need not be locked together. This eliminates the added circuit or extra communications channel needed to synchronize the two clocks. The tolerance of the clock rates in asynchronous transmission is said to be 0.01 percent, but this is easily obtainable without too much trouble.

The UART is a single large-scale integration chip that will perform all of the data transmission functions on the digital side. It is programmable as to bit length, parity bits, and the overall length of the stop bits.

The block diagram to a common UART IC is shown in Fig. 14-11. This particular device is the TR1602A/B by Western Digital. Note that the pinouts for the UART are almost universally standardized and are based on the now obsolete AY-1013. Most UARTs are capable of all three communications modes: simplex, half-duplex and full-duplex. This feature is due to the fact that all

Fig. 14-10. Transmit connections for the model 33.

Fig. 14-11. Block diagram of a UART.

of the transmitter and receiver control pins are independent of each other.

The UART is capable of being user-programmed as to transmitted word length, baud rate, parity type (odd-even, receiver verification/transmitter generation), parity inhibit, and stop bit length (1, 1.5, or 2 clock periods). The UART also provides six different status flags: transmission completed, buffer register

194

transfer completed, received data available, parity error, framing error, and overrun error.

The maximum clock speed is between 320 kHz and 800 kHz, depending upon the particular type selected. Note that the clock rate actually used in any given application is dependent upon the baud rate. The clock frequency is 16 times the data baud rate.

The receiver output lines are tri-state, which means there is a high-impedance to both ground and V+ when the outputs are inactive. This allows the outputs of the receiver section to be connected directly to a data bus without extra circuitry.

The transmitter section uses an eight-bit input register. This makes it capable of accepting an eight-bit parallel word from a source such as a keyboard, computer output port, etc. It will assemble these bits and then transmit them at the designated time, adding any demanded parity or stop bits.

The receiver data format is a logical mirror image of the transmitter section. It will input serial data bits, strip off the start, stop and parity bits (if used), and then assemble the binary word in parallel form. In addition, it will test the data for validity by comparison of the parity bits and stop bits.

The standard UART data format is shown in Fig. 14-12. The data line (transmitter serial output or receiver serial input) will normally sit at a logical HIGH level unless data is being transmitted or received. The start bit (B0) is always LOW, so the HIGH to LOW transition is what the UART senses for the starting of a word or transmission. Bits B1 through B8 are the data bits loaded into the transmitter register from the outside world. Although Fig. 14-12 shows all eight bits, the user can program the device for fewer if needed. Bit lengths of 5, 6, 7, or 8 bits are allowed. Bits are dropped in shorter formats from the B1 side of the chain.

Figure 14-13 shows typical receiver and transmitter con-

Fig. 14-12. Serial data format from UART.

Fig. 14-13. Transmitter connections on UART at A, and receiver connections on UART at B.

196

figurations for the UART. The transmitter section is shown in Fig. 14-13A, while Fig. 14-13B shows the circuitry for a receiver section. If the serial output of the transmitter is connected to the serial input of the receiver, a closed loop exists. The output word from the receiver would match the transmitted word. In most cases, the UART is used to drive some external communications channel, such as an audio-frequency shift keyer, for transmission over some standard communications media.

Neither digital levels representing the two logic states nor binary bits can be applied directly to telephone lines or two radio communications channels. Most common radio and telephone channels want tones in the 300 to 3000 Hz range with some extended bandwidth facilities wanting to see 100 to 10,000 Hz. Before bits can be transmitted along one of these media, they first must be converted to audio tones that are compatible with the telephone lines or radio modulator input. Audio-frequency shift keying (AFSK), similar to that used in some forms of teletypewriter communications, is generally used to represent the two tones.

Various standards have been developed that assign tone pairs (one each for logical-1 and logical-0) that may be used. Again, this is done for the sake of standardization throughout the industry. In some cases, manufacturers will use the Bell System's *Dataset*® tones; in others, they will use one of the other actual or de facto standards. Some use the regular telephone touch-tone frequencies, in which *two* frequencies are used to represent each of 10 digits plus several control signals. Each character is

Table 14-2. Touch-Tone Frequencies.

Digit	Tone	Frequencies
0	941 and	1336 Hz
1	697	1209
2	697	1336
3	697	1477
4	770	1209
5	770	1336
6	770	1477
7	852	1209
8	852	1336
9	852	1477
*	941	1209
#	941	1477
	(Low group)	(High group)

represented by a high frequency and a low frequency. Table 14-2 lists the several touch-tone frequencies.

Touch-tone signals are most easily generated using a specialized integrated circuit, such as the Motorola MC14410. This chip is an LSI CMOS device that is designed to generate the two frequencies required on command from a standard two-of-eight contact closure keypad. Several manufacturers sell these pads in a standard format. They are arranged in a 4 × 4 matrix: four rows and four columns. When a pushbutton is depressed, a crossbar switch shorts together one row and one column, creating a unique logic combination out of a possible 16.

The MC14410 contains its own timing circuits, including a clock oscillator that operates off a standard 1000-kHz crystal (external). The chip creates two semi-sine waves, digitally using a special addition technique. The tolerance of the output frequency is on the order of 0.2 percent, well within the touch-tone specification.

The basic connection scheme for the MC14410 is shown in Fig. 14-14. Note that very little in the way of external circuitry is required on this chip; it is almost self-contained. The only connections required in this most basic configuration are four row

Fig. 14-14. Connection of the MC14410 Touch-tone generator.

Fig. 14-15. Tone encoder for binary data.

inputs, four column inputs, ground, +5 VDC, and the high-frequency and low-frequency output terminals. Note that the two outputs are independent of each other, allowing them to be used separately, if needed. In most applications, however—especially actual telephone circuits—these outputs are summed in a resistor mixing network.

The keyboard inputs to the MC14410 want to see a LOW to ground in order to be selected. The keyboard will simultaneously, in crosspoint manner, enable only the two inputs that correspond to the number or character on that key top.

An example of a frequency shift keyer is shown in Fig. 14-15. In this circuit, the two frequencies (F1 and F2) are generated in clock oscillator circuits or in a countdown circuit from a master clock (as appropriate). Two TTL NAND gates are used to control the output frequency. F1 is applied to one input of G1, while F2 is applied to one input of gate G2. The data line is the binary bit that is being represented; i.e., it is HIGH for a logical-1 and LOW for a logical-0. When the data line is LOW, then gate G2 is inhibited; according to the rules for a NAND gate, it will have a HIGH output if either input is LOW. Because of the inverter, however, the input of gate G1 is HIGH. Its output line, therefore, is allowed to follow the F1 clock. Similarly, when the data input is HIGH, gate G1 is inhibited, and G2 follows the F2 clock. Clearly, then, F1 is output when the data line is LOW, and F2 is output when the data line is HIGH.

Figure 14-15 shows a resistor mixing network. This will sum the two outputs and reduce their amplitude, so that they can be applied to the same communications circuit. Most communications equipment that requires audio input signals will not tolerate TTL level signals, so the amplitude must be reduced. If the communications channel specifically asks to see a TTL logic level at the input, though, then replace the resistor network with a two-input NOR gate (one input tied to the G1 output and the other to the G2 output).

Several other circuits are commonly used, including a few special-purpose IC devices. Rarely will you find something like a voltage-controlled oscillator or an oscillator that is started up when the appropriate logic level is applied. This is too messy and tends to pull the oscillator a little when first turned on. The resultant output is not unlike chirp in CW radio transmitters.

Although circuits differ somewhat from one manufacturer to another, there are two basic forms of tone decoder circuits: filters and phase-locked loops (PLL). An example of the *filter* style of decoder circuit is shown in Fig. 14-16. An input amplifier may or may not be used, depending upon the circumstances. If the signal is weak and noisy, as it might well be after transmission over long-distance telephone lines, the amplifier is needed. The bandwidth of the amplifier is narrow enough to pass only the two frequencies or any additional frequencies used in control applications. This eliminates much of the noise that comes in with the signal. The output of the amplifier is applied to the inputs of two very narrow-band, high-Q, band-pass filters. One filter will pass only F1, while the other will pass only F2. The outputs, of the two filters are

Fig. 14-16. Block diagram of tone decoder using filter method.

Fig. 14-17. PLL tone decoder.

usually then rectified, and the rectified—but not smoothed—signal is applied to one input of a voltage comparator. Recall that a comparator circuit will produce an output that indicates whether or not the input voltages are in agreement. If F1 is present, for example, the comparator would be designed to produce a LOW output, but if F2 is seen, then a HIGH is output. In general, the comparator method is probably the most commonly used in low-speed data communications applications.

But the phase-locked loop of Fig. 14-17 is not without supporters. A phase-locked loop is a circuit that will adjust the frequency of a VCO to keep it in sync with an input signal. In a tone decoder, the input signal is the desired tone. In the circuit of Fig. 14-17, the output (pin No. 8) goes LOW when the desired tone is applied to the input. We can use two of these circuits—one for each tone—to decode AFSK signals used in data communications with a little extra logic circuitry. For the most part, however, this type of circuit is used in control applications where the tone controls a computer, other processor, or radio repeater. Again by the use of appropriate external logic circuits, the entire range of 16 possible input signals can be decoded into 1-of-16 outputs. The response frequency of this circuit is given approximately by:

$$F = \frac{1}{(R1 + R2)\, C1}$$

where F is the frequency in hertz, R1 and R2 are in ohms, and C1 is in farads.

TONE MULTIPLEXING

Chapter 13 covered time domain multiplexing and which signals were sampled and then transmitted over a common media by interleaving the samples. If the timing is carefully controlled and synchronized at both ends, the total sampled data signals are recovered at the receiving end of the system. But time domain multiplexing, or TDM, is not the only form of multiplexing available. It is, however, preferred in *local* systems. Once we go outside of the immediate system and transmit data over radio or telephone facilities, some form of frequency domain multiplexing (FDM) is usually specified. Of course, high-density data communications systems may well use a combination of *both* TDM and FDM to greatly increase the bit rate (at, incidentally, the cost of a much greater required bandwidth).

A simplified version of FDM is shown in Fig. 14-18. In Fig. 14-18A, both channels 1 and 2 occupy the same baseband. They are both ordinary 300 - 3000 hertz audio communications channels. If these were transmitted over the same set of telephone lines, they would interfere with each other immensely. But if we can translate them to different frequencies, both can be sent over the same set of lines. Filters at the receiver end will recover them and split them off. In Fig. 14-18B, left channel No. 1 is in the baseband (300 to 3000 Hz) but shifted by heterodyning channel No. 2 into the range 3300 to 6000 Hz. The bandwidth occupied by channel No. 2 is the same 2700-Hz as it had when it was in baseband. But the upper and lower limits have been changed from 300 to 3000 Hz to 3300 to 6000 Hz.

In actual practice, audio grade lines can handle three—maybe

Fig. 14-18. Two channels of data occupy the same baseband at A, and frequency domain multiplexing of the second channel at B.

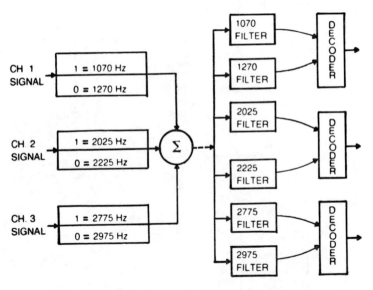

Fig. 14-19. Multi-channel frequency domain multiplexer.

four—shifted channels. Video grade lines, of course, can handle much more. But these figures are for voice communications channels or wideband data channels. Ordinary slow-speed to medium-speed data channels can use only two frequencies for each data bit (HIGH/LOW = F2/F1). While the actual bandwidth required will be determined by the speed of transmission, many more AFSK channels will fit into the same spectrum space occupied by only two or three voice channels.

Figure 14-19 shows a typical three-line data communications system based on FDM principles. Each channel uses a tone pair in which F1 and F2 are approximately 200-Hz apart. Channel No. 1 uses 1070 Hz for HIGH and 1270 Hz for LOW. Channel No. 2 uses 2025 Hz for HIGH and 2225 Hz for LOW, while Channel No. 3 uses 2775 Hz for HIGH and 2975 Hz for LOW. Notice that we have inserted three, one-bit, data channels in the single-channel baseband space needed by voice communications. The tone outputs of all three encoders are applied to a combiner, usually a resistive network, and transmitted along the telephone or radio communications channel. At the receive end of the path, the signal composite is applied to the inputs of some high-Q filters. In some cases, a pre-filter will exist for each channel that will pass only the two desired frequencies. From then on, one of the decoder circuits

Fig. 14-20. Equivalent circuit of the telephone system at A, and voltage on and off hook at B.

Fig. 14-21. Telephone patch.

shown previously will output a HIGH or LOW, depending upon which is required.

TELEPHONE LINES

The commercial telephone lines can now be accessed by non-ATT users. While it is necessary to subscribe and have the local telephone company install the required grade of line, it is no longer necessary for you to buy telephone company's equipment. For ordinary voice-grade lines, the bandwidth will be 300 to 3000 Hz. The transmitting source is expected to output approximately -5 VU of single (VU is decibel scale referenced such that 0 VU is 1 mW in a 600-ohm load). The receiver is expected to tolerate signals in the -45 VU to -10 VU range.

Telephone lines tend to be long, so their ohmic resistance is significant. Fig. 14-20A shows an equivalent circuit, neglecting capacitance and inductance effects. The receiver looks like a resistance on the end of the line, while the R_{line} is the cumulative series resistance of the telephone line. When the receiver is on the hook, its internal switch is open, so the voltage (see Fig. 14-20B) is approximately 60V. But when the receiver is lifted off the hook, the switch is closed and R_{rcvr} is in the circuit. This will load the circuit considerably. The voltage across the telephone receiver is now:

$$E = \frac{(60V)\ (R_{rcvr})}{(R_{line}\ +\ R_{rcvr})}$$

In most cases, it will be in the 6V to 26V range.

Figure 14-21 shows a method used in radio stations, police communications, and some amateur radio stations to connect the telephone line to receivers and transmitters. The same principle is also used in data communications using AFSK techniques. This circuit is called a phone patch. It is essentially a bridge that mixes the audio from the transmitter and the audio to the receiver in a linear but compensated network. The telephone line is one leg of the bridge. Note the use of the voice coupler. This is a protective device supplied by the telephone company. In the past, it was always necessary to rent a voice coupler from "Ma Bell." But it has recently become possible to buy one from a commercial outfit, provided that they guarantee it to meet telephone company specifications. Also, some local telephone companies have opened telephone stores that sell such accessories.

Table 14-3. UART Pinouts.

Pin No.	Mnemonic	Function
1	Vcc	+5 volts DC power supply.
2	VEE	−12 volts DC power supply.
3	GND	Ground.
4	RRD	Receiver Register Disconnect. A high on this pin disconnects (i.e., places at high impedance) the receiver data output pins (5 through 12). A low on this pin connects the receiver data output lines to output pins 5 through 12.
5	RB8	LSB
6	RB7	
7	RB6	
8	RB5	Receiver data output lines
9	RB4	
10	RB3	
11	RB2	
12	RB1	MSB
13	PE	Parity error. A high on this pin indicates that the parity of the received data does not match the parity programmed at pin 39.
14	FE	Framing Error. A high on this line indicates that no valid stop bits were received
15	OE	Overrun Error. A high on this pin indicates that an overrun condition has occurred, which is defined as not having the DR flag (pin 19) reset before the next character is received by the internal receiver holding register.
16	SFD	Status Flag Disconnect. A high on this pin will disconnect (i.e., set to high impedance) the PE, FE, OE, DR, and THRE status flags. This feature allows the status flags from several UARTs to be bus-connected together.
17	RRC	16 × Receiver Clock. A clock signal is applied to this pin, and should have a frequency that is 16 times the desired baud rate (i.e., for 110 baud standard it is 16 × 110 baud, or 1760 hertz).
18	DRR	Data Receive Reset. Bringing this line low resets the data received (DR, pin 19) flag.
19	DR	Data Received. A high on this pin indicates that the entire character is received, and is in the receiver holding register.
20	RI	Receiver Serial Input. All serial input data bits are applied to this pin. Pin 20 must be forced high when no data is being received.
21	MR	Master Reset. A short pulse (i.e., a strobe pulse) applied to this pin will reset (i.e., force low) both receiver and transmitter registers, as well as the FE, OE, PE, and DRR flags. It also sets the TRO, THRE, and TRE flags (i.e, makes them high).
22	THRE	Transmitter Holding Register Empty. A high on this pin means that the data in the transmitter input buffer has been transferred to the transmitter register, and allows a new character to be loaded.
23	THRL	Transmitter Holding Register Load. A low applied to this pin enters the word applied to TB1 through TB8 (pins 26 through 33, respectively) into the transmitter holding register (THR). A positive-going level applied to this pin transfers the contents of the THR into the transmit register (TR), unless the TR is currently sending the previous word. When the transmission is finished the THR→TR transfer will take place automatically even if the pin 25 level transition is completed.
24	TRE	Transmit Register Empty. Remains high unless a transmission is taking place, in which case the TRE pin drops low.
25	TRO	Transmitter (Serial) Output. All data and control bits in the transmit register are output on this line. The TRO terminal stays high when no transmission is taking place, so the beginning of a transmission is always indicated by the first negative-going transition of the TRO terminal.
26	TB8	LSB
27	TB7	
28	TB6	
29	TB5	Transmitter input word.
30	TB4	
31	TB3	
32	TB2	
33	TB1	MSB

34	CRL	Control Register Load. Can be either wired permanently high, or be strobed with a positive-going pulse. It loads the programmed instructions (i.e., WLS1, WLS2, EPE, PI, and SBS) into the internal control register. Hard wiring of this terminal is preferred if these parameter never change, while switch or program control is preferred if the parameters do occassionally change.
35	PI	Parity inhibit. A high on this pin disables parity generation/verification functions, and forces PE (pin 13) to a low logic condition.
36	SBS	Stop Bit(s) Select. Programs the number of stop bits that are added to the data word output. A high on SBS causes the UART to send two stop bits if the word length format is 6, 7, or 8 bits, and 1.5 stop bits if the 5-bit teletypewriter format is selected (on pins 37-38). A low on SBS causes the UART to generate only one stop bit.
37 38	WLS1 WLS2 }	Word Length Select. Selects character length, exclusive of parity bits, according to the rules given in the chart below:

Word Length	WLS1	WLS2
5 bits	low	low
6 bits	high	low
7 bits	low	high
8 bits	high	high

| 39 | EPE | Even Parity Enable. A high applied to this line selects even parity, while a low applied to this line selects odd parity. |
| 40 | TRC | 16 × Transmit Clock. Apply a clock signal with a frequency that is equal to 16 times the desired baud rate. If the transmitter and receiver sections operate at the same speed (usually the case), then strap together TRC and RRC terminals so that the/same clock serves both sections. |

Fig. 14-22. MC14408 and MC14409 telephone dial pulser.

207

One last device to be shown, is the telephone dialer IC of Fig. 14-22. This is a special Motorola chip (MC14408 or 14409) that interfaces digital contact closure inputs with dial telephone circuits. Both this IC and the touch-tone MC14410 can be directly controlled by a digital circuit or through a manual keyboard. Most of the automatic dialers will use such a circuit—usually the MC14410—and then output the desired numbers stored in binary registers.

Chapter 15

Clocks: Oscillators And Multivibrators

Oscillators and multivibrators are used in many different types of electronic circuits, including digital circuits and computers. This chapter is a broadly-based treatment of oscillators and multivibrators. These circuits are used to produce electronic waveforms that are needed to synchronize the different circuits of the computer. While many of the circuits shown here are used only marginally in computers and somewhat more in other digital circuits, all of them operate on similar principles, so they are covered in this chapter. You may want to use this chapter in a more broadly based context, and keep it handy as a reference.

In general, let's keep our jargon straight by adopting the term "oscillator" to cover all cases, including astable multivibrators and digital clocks. The term oscillator can be defined as a circuit that will produce a periodic waveform (i.e., one that repeats itself). The periodicity of the waveform makes it a clock in the digital context. The output waveform can be a sinewave, squarewave, triangle, sawtooth, pulse, or other waveshape, but before it is used in a computer it must be converted to either squarewave or trapezoidal waveshapes.

There are two forms of oscillator circuit: *relaxation oscillators* and *feedback oscillators*. Some relaxation oscillators use one or another of several "negative resistance" devices (e.g., tunnel diode). Such devices operate according to Ohm's law at certain potentials, and the opposite of Ohm's law at other potentials. There are other relaxation oscillators that use electronic devices which pass little or no current at voltages below some threshold, and pass a large current at voltages above the threshold. Examples of these devices are neon glow-lamps and unijunction transistors (UJT).

Feedback oscillators use an active device as an amplifier, and then provide feedback in a manner that produces regeneration instead of degeneration (i.e., positive feedback instead of negative feedback). These circuits account for a large number of oscillators used in practical electronic circuits. In this series of articles, we will discuss both forms of oscillator.

RELAXATION OSCILLATORS

There are two main devices used in practical relaxation oscillators: neon glow-lamps and UJTs. There are also some circuits that are based on SCR-like devices, but these are not common.

Figure 15-1A shows a simple relaxation oscillator based on the neon glow lamp. The glow lamp has a low pressure inert gas (neon) inside of a glass envelope that also contains a pair of electrodes. When the potential across the electrodes exceeds the ionization potential (V_t in Fig. 15-1B) of the inert gas, the gas will give off light. Popular types of neon glow lamp include the NE-2 and the NE-51. For these lamps, the threshold potential is somewhere between 40 and 80 volts, although the lamp will maintain its ionized state (and continue to glow) at a somewhat lower "holding potential" (V_H).

The circuit for the relaxation oscillator is shown in Fig. 15-1A, while its operating waveforms are shown in Fig. 15-1B. The frequency of oscillation is set by the ionization and holding potentials, acting in concert with the RC time-constant of the resistor and capacitor. The resistor is connected in series with the lamp, while the capacitor is connected in parallel with the lamp.

We must keep in mind the two states of the neon lamp. When the lamp is not ionized, it conducts no current. This is the situation at all potentials below V_t. At voltages above the threshold, the lamp resistance suddenly drops to nearly nothing (and in fact, is so high that it will blow the lamp if the resistor value is too low). When the circuit is turned on, the capacitor will begin to charge and the capacitor voltage (V_c) will begin to build up. When the voltage across the capacitor reaches the threshold potential, the lamp ionizes and takes on a very low resistance. Since this low resistance is across the capacitor, it will discharge the capacitor. But the discharge only continues until V_c goes below the threshold

Fig. 15-1. A) Glow lamp relaxation oscillator, B) timing waveform.

potential. When this potential is reached, the lamp goes out and reverts to its high resistance state. The cycle continues, with the capacitor voltage varying between threshold and holding potentials. The frequency of oscillation is determined by the difference between these potentials and the values of R and C.

Unijunction Transistors. The UJT is a special form of transistor that has one emitter, two bases and no collector (see Fig. 15-2). The structure of the UJT is shown in Fig. 15-2A, while the circuit symbol is shown in Fig. 15-2B. The UJT is constructed of a single channel of silicon, and the "bases" are merely the electrodes at either end of the channel. The emitter forms a PN junction with the channel. Like most PN junctions, the UJT PN junction will not pass current when reverse or zero biased, but will pass a current when forward biased (usually 0.6 to 0.7 volts is required to forward bias the PN junction).

Fig. 15-2. A) Unijunction transistor, B) circuit symbol.

An example of a UJT relaxation oscillator is shown in Fig. 15-3. The frequency of oscillation is set by the combination R1 and C1, acting together with the properties of the UJT. When the circuit is turned on, the B1-E PN junction is unbiased so there will be no current flowing in resistor R3. At power on, the capacitor will begin to charge from V+ through timing resistor R1. When the threshold potential is crossed, the UJT will "fire" and a pulse of current flows in R3. This produces (by Ohm's law) a pulse output from across R3—this is the output signal.

Negative Resistance Devices. There are several semiconductor devices on the market that have a so-called negative resistance property. What does this term mean? Regular, or "positive resistance," devices operate according to Ohm's law. When the voltage across the circuit is increased, the current in the device increases proportionally; when the voltage is decreased, the current decreases also. In a negative resistance device, however, exactly the opposite is found. An increase in voltage will produce a decrease in current, and vice versa.

An example of a negative resistance device is the tunnel diode (also called an Esaki diode after the inventor), the symbols for which are shown in Fig. 15-4A. The I-vs-V curve for the tunnel diode is shown in Fig. 15-4B. In the positive resistance zone (PRZ), the tunnel diode operates much like other "Ohm's law" devices. But in the negative resistance zone (NRZ), it behaves just the opposite from Ohm's law!

An example of a tunnel diode oscillator is shown in Fig. 15-5. This particular circuit was popular in the mid '60s for small Citizen's Band beacons and hobby oscillators, and may reappear in ultrafast

212

Fig. 15-3. Unijunction transistor relaxation oscillator.

Fig. 15-4. A) Tunnel diode symbols, B) I-vs-V characteristic.

computers. With the circuit constants shown, the oscillator will operate in the 27 MHz region using an overtone crystal. Other circuits, similar in concept but different in form, can cause a tunnel diode to operate into the UHF or even microwave region. In those

Fig. 15-5. Tunnel diode oscillator.

214

cases, strip-line tuners and cavities are used to establish resonant frequency.

Relaxation oscillators are simple, and are good circuits for students and beginning hobbiests to build (if only to learn how they work). Most other applications, however, will use one or another feedback oscillator circuits.

FEEDBACK OSCILLATORS

A feedback oscillator consists of an amplifier and a feedback network (see Fig. 15-6). As a matter of interest let's see if we can derive the gain equation for Fig. 15-6. Assume that A_{vol} is the open-loop (i.e., no feedback) gain of the amplifying element, and A_v is the gain of the overall circuit when the feedback circuit is connected. Further, we will define the following: amplifier output voltage is V_o, input signal voltage is V_{in}, the feedback network output voltage is V_f, and the difference between input and feedback signal voltages is V_d. The gain (or "transfer function") of any circuit is simply the ratio of the output over input voltages. With this information we can conclude the following:

$$B = V_f/V_o \qquad \text{Eq.(1)}$$

$$A_{vol} = V_o/V_d \qquad \text{Eq.(2)}$$

$$A_v = V_o/V_{in} \qquad \text{Eq.(3)}$$

We can rearrange Eq.(2) to find V_o:

$$V_o = V_d A_{vol} \qquad \text{Eq.(4)}$$

Fig. 15-6. Feedback oscillator.

215

And then substitute Eq.(4) into Eq.(3):

$$A_v = V_o/V_{in} = V_d A_{vol}/V_{in} \qquad \text{Eq.(5)}$$

We know that:

$$V_d = V_{in} - V_f \qquad \text{Eq.(6)}$$

So,

$$V_{in} = V_d + V_f \qquad \text{Eq.(7)}$$

We can now use Eq.(7) in Eq.(5):

$$A_v = \frac{V_d A_{vol}}{V_d + V_f} \qquad \text{Eq.(8)}$$

Rearranging Eq.(1) yields the expression for V_f:

$$B = V_f/V_o \qquad \text{Eq.(9)}$$

So,

$$V_f = V_o B \qquad \text{Eq.(10)}$$

Substituting Eq.(10) into Eq.(8):

$$A_v = \frac{V_d A_{vol}}{V_d + V_f} = \frac{V_d V_{ol}}{V_d + V_o B} \qquad \text{Eq.(11)}$$

Our job is nearly complete. We need to find an expression of V_o in terms of V_d; Eq.(4) fills that need, so by substituting Eq.(4) into Eq.(11):

$$A_v = \frac{V_d A_{vol}}{V_d + V_o B} = \frac{V_d A_{vol}}{V_d + (V_d A_{vol})B}$$

Since V_d is in both numerator and denominator, we can divide it out of Eq.(12), leaving:

$$A_v = \frac{A_{vol}}{1 + A_{vol}B}$$

216

The *loop gain* of the circuit is the product of the feedback factor (B) and the open-loop gain (A_{vol}):

$$LOOPGAIN = -A_{vol}B$$

The amplifier can be any of many different devices. For some cases, it will be a common-emitter bipolar transistor (NPN or PNP devices). In others it will be a junction field effect transistor (JFET) or metal oxide semiconductor field effect transistor (MOSFET). In older equipment it was a vacuum tube, most often operated in the common cathode mode. In still other cases, it will be an integrated circuit operational amplifier, or some other form of IC amplifier. In most cases, the amplifier will be an inverting type, so that the output is out of phase with the input by 180 degrees.

The feedback network will provide a sample of the output signal to the input of the amplifier. This network may be LC, RC, or RLC. Rarely is it a simple resistance network in oscillator circuits, however, because such networks are not frequency selective. They will produce oscillation, however, but at some frequency that is not controlled. The problem is that there are distributed ("stray") circuit capacitances in the circuit, and these act with the resistance to produce an RC time constant at some unspecified frequency, which is rarely a frequency that you want!

This circuit will oscillate if two conditions are met: 1) the loop gain (as above) is unity, and 2) the feedback signal is in phase with the input signal. These conditions are called the "Barkhausen criteria". Since the typical amplifier inverts the signal (180 degrees), and since we need a total phase shift of 360 degrees (in order to have an in-phase feedback signal), the feedback network must provide an additional phase shift of 180 degrees. If the network is designed to produce this required phase shift at only one frequency (as in an LC or RC network), then the circuit will oscillate at that frequency.

Earlier we determined that there are two basic forms of oscillator circuit: *relaxation* and *feedback*. The relaxation oscillator depends upon the properties of certain devices such as the neon glow lamp and the unijunction transistor. Such devices will present a high impedance at applied potentials (usually the voltage across a timing capacitor) less than a certain threshold. Once the threshold is passed, however, the device resistance breaks down to a low value and this change discharges the capacitor. Feedback oscillators, on the other hand, depend upon applying a sample of the output signal of an amplifier to its own input in-phase, thus causing regeneration rather than degeneration.

Figure 15-6 was the block diagram for the feedback oscillator. The forward block is an amplifier element that has an open-loop gain of $-A_{vol}$ (the minus sign implies 180-degree phase inversion between input and output). The feedback network has a "gain" designated by "B", and may consist of a simple resistance network, or a complex impedance. In most cases, its "gain" is really a loss. The value of the closed loop gain (which takes into account both A_{vol} and B) was defined in earlier and is:

$$A_v = \frac{A_{vol}}{1 + A_{vol}B}$$

If this circuit is to oscillate, then we must make the closed loop gain equal to or greater than unity. It is also necessary to ensure that the feedback signal is in phase with the input. Since the amplifier has a 180-degree phase shift, the feedback network must provide an additional phase shift of 180 degrees at the frequency of oscillation. The combination of phase shifts produces the 360-degree phase shift needed for the in-phase condition at one frequency.

Figure 15-7 shows a block diagram of a typical LC oscillator circuit. The amplifying element is an operational amplifier, but it could be any form of inverting voltage amplifier. The use of the op-amp here is for convenience of illustration rather than an endorsement.

In this general case, we designate the reactance elements with the general terms Z1, Z2, and Z3, meaning that they can be either capacitive or inductive reactances. The feedback factor, B, is given by the standard "voltage divider equation" (Fig. 15-7 shows that it is similar to a resistor voltage divider):

$$B = \frac{Z2}{Z2 + Z3}$$

The usual arrangement is to use the same form of reactance for Z1 and Z2, and then the opposite form for Z3. For example, if Z1 and Z2 are capacitors, then Z3 is an inductor; if Z1 and Z2 are inductors then Z3 is a capacitor.

Fig. 15-7. Block diagram of an LC oscillator.

LC OSCILLATORS

Let's digress a moment to look at LC (inductor/capacitor) resonant tank circuits. Figure 15-8A shows an LC tank circuit in a test circuit. When switch S1 is closed, current will flow and energy is stored in the inductor. But when S1 is opened, the magnetic field of the inductor collapses, inducing a current into the circuit that tends to charge capacitor C. When C is charged, the process reverses, the capacitor discharges and reforms a magnetic field in the inductor. But this time, circuit losses have reduced the amount of available energy, so the field is less strong than before. The fields alternate back and forth between capacitor and resistor, each one being a little less strong that its predecessor—with the result being the damped waveform of Fig. 15-8B. Each time the LC tank circuit is "pulsed" with a bit of energy by closing S1, the oscillations will begin and have a frequency given by the standard resonance equation:

$$F = \frac{1}{2\pi \sqrt{LC}}$$

COLPITTS OSCILLATORS

The circuit for the Colpitts LC oscillator is shown in Fig. 15-9.

Fig. 15-8. A) LC tank circuit, B) damped oscillation.

In this case, we see that reactances Z1 and Z2 are capacitors C1 and C2, respectively, while reactance Z3 is represented by inductor L1. This scheme is based on the rule of thumb given above of making Z1 and Z2 the same form of reactance, while Z3 is the opposite form of reactance. The operating frequency is given by the standard resonance formula (as in the equation above) with the terms redesignated for the specific circuit:

$$F = \frac{1}{2\pi \sqrt{L1 \; C}}$$

Where:

L1 is the inductance in Henrys

C is the capacitance of the combination C1 and C2, in farads (see below).

The unit "C" is defined as the series combination of C1 and C2, which is found from the standard expression:

$$C = \frac{C1 \times C2}{C1 + C2}$$

The minimum gain of the circuit is the ratio C1/C2, and we must make C2 greater than C1 if the gain is to be less than unity. In actual practice we find that the ratio of C1/C2 is selected as a trade-off between amplitude and stability.

Hartley Oscillators

The classical Hartley oscillator is shown in Fig. 15-10. The principal difference between the Hartley and Colpitts oscillators is the arrangement of Z1, Z2, and Z3. In the Colpitts oscillator, Z1 and Z2 were capacitive, while in the Hartley they are inductive. In other configurations of these circuits, we recognize a Hartley oscillator by a tapped inductor, and the Colpitts by a capacitive voltage divider. In the circuit of Fig. 15-10 the frequency of oscillation is:

$$F = \frac{1}{2\pi \sqrt{(L1 + L2)C1}}$$

Fig. 15-9. Colpitts oscillator.

Fig. 15-10. Hartley oscillator.

The minimum gain is L2/L1, while the actual gain is R2/R1.

Practical Colpitts and Hartley Circuits

The circuits shown in Figs. 15-9 and 15-10 are models, while practical versions are shown in Fig. 15-11. An example of the Colpitts oscillator is shown in Fig. 15-11A. The split capacitor voltage divider that provides feedback consists of C1 and C2; in general, C2 is larger than C1 (typical values are 82-pF for C1 and 0.001-μF for C2). Resistor R3 is used for biasing the transistor, while R1 provides a little stability; the output signal is developed across R1. The frequency of oscillation is provided by a parallel resonant tank circuit.

A variation on the circuit, called the "Clapp oscillator" is the same as the Colpitts, except that the LC tank circuit is series resonant (see Fig. 15-11A).

The Hartley version is shown in Fig. 15-11B. Again, a bipolar transistor is used for gain purposes. Resistors R1 and R2 are used for biasing the transistor, but play no other role. The output signal

Fig. 15-11. A) Transistor colpitts oscillator, B) transistor Hartley oscillator.

can be taken either from the emitter of the transistor, or from a separate link-coupled coil to the main frequency-setting inductor.

The resonant frequency is set by the capacitor (C1) and the total inductance of L1A and L1B. The setting of the tap that divides L1A from L1B is a trade-off between stability and output amplitude.

Both Hartley and Colpitts/Clapp versions are common collector circuits. In both cases, the collector of the transistor is set to AC ground by using a bypass capacitor (C4 in Fig. 15-11B, and C5 in Fig. 15-11A).

Other LC Oscillators

There are at least two additional forms of LC oscillator, and both of them are among the oldest forms of electronic oscillator: the Armstrong oscillator is shown in Fig. 15-12, while the Tuned-Base-Tuned-Collector Oscillator (TBTC) is in Fig. 15-13.

The Armstrong oscillator is named after Major Edwin Armstrong, who invented the regenerative detector, superheterodyne radio and frequency modulation, among other things. In another form, the Armstrong oscillator becomes the regenerative radio detector.

The frequency of oscillation for the Armstrong oscillator is set by the parallel resonant tank circuit C1L1. A feedback "tickler" coil (L2) is closely coupled with L1, and serves to feedback signal from the output to the input (drain to gate, in the case of a field

Fig. 15-12. Armstrong oscillator.

Fig. 15-13. Tuned input/tuned output oscillator.

effect transistor). The "sense" (phase) in which L2 is connected will determine whether the feedback is positive or negative.

Regeneration, or oscillation if you will, is controlled by a series resistor that controls the level of current flowing in the tickler coil. In older circuits the feedback control is mechanical. In those circuits, the tickler coil is on a moving mechanism that permits it to be positioned relative to L1. When the coil is perfectly aligned, the coupling is maximum (and so is the feedback). The frequency of oscillation is determined by the same resonance equations used in examples above in which L1 and C1 are the inductive and capacitive elements, respectively.

Figure 15-13 shows the TBTC oscillator. In older texts, which used vacuum tube examples, this circuit was called the Tuned-Grid-Tuned-Plate (TGTP) oscillators. Perhaps, in this day when we have many forms of amplifying device, this circuit should be renamed the "Tuned-Input-Tuned-Output (TITO) oscillator." The oscillating frequency is set by both L1C1 and L2C2. Most textbooks recommend that these two resonant tank circuits be at slightly different, but almost the same, resonant frequencies.

Feedback in the TITO oscillator is determined by the

225

interelement capacitances of the transistor. These capacitances are not physically in the circuit, but are inherent in the devices: C_{be} is the capacitance between base and emitter of Q1, while C_{cb} is the capacitance between collector and base.

The TITO oscillator is rarely used these days, even though some VHF and UHF TV and FM receiver local oscillators are implicitly TITO. In those cases, the capacitance may not be immediately obvious and the circuit will appear to have only inductance. The designers intended that the sum of the stray capacitances in the circuit and the input capacitances of the tube or transistor used as the local oscillator will resonate the coil selected.

At this point it may be well to ask how oscillation begins? After all, from an intuitive point of view, it may seem that the oscillator lacks an initial stimulus to get things started. If the oscillation depends upon feeding back the output signal to the input in-phase, then how on earth does the process begin initially when there is no output signal to feedback? In mechanical analogies to oscillators, such as musical tuning forks, we need a mechanical stimulus—is there an electrical equivalent to rapping the tuning fork that gets the process going?

The answer, of course, is yes. We do, in fact, have an initial stimulus. First, there is always noise in any electrical circuit—none are every perfectly noise free. In fact, active devices such as transistors and tubes seem to have an abundance of noise that initially "rings" the circuit. Secondly, there is also a varying output potential caused by the changing collector or drain current in the transistor as the device "comes alive". In many cases this delta-I is sufficient to ring the circuit into oscillation. If the circuit were too quiet or too slow in turning on, then there would be no oscillations.

RC OSCILLATORS

Resistor and capacitor elements can be used to time an oscillator or clock circuit. In fact, in most low-cost computers the master clock circuit is an RC oscillator. We will look at several different types in order to learn the principles.

RC Phase-Shift Oscillator

The basic RC phase shift oscillator uses a three-section RC network to provide 180 degrees of phase shift. When added to the

180 degrees provided by an inverting amplifier, this phase shift meets the 360-degree criteria required for oscillation. Figure 15-14 shows two basic forms of RC phase shift oscillators. The circuit in Fig. 15-14A is based on a junction field effect transistor (JFET) amplifier, while the circuit in Fig. 15-14B is based on the operational amplifier. Both of these circuits produce a sinewave output signal.

In both cases, there is a three-section phase shift network: R1/C1, R2/C2, and R3/C3. These three networks are all identical to each other, and are designed to provide 60-degrees of phase shift at the desired oscillating frequency. When the three phase shifts are added together, we find the required 180-degree phase shift. In this example, it is assumed that R1 = R2 = R3 = R, and C1 = C2 = C3 = C (where R and C are the resistance and capacitance values used in the equations). As in most cases, we will use the unit farads for C, ohms for R and hertz for the frequency. The oscillating frequency is given by:

$$F = \frac{1}{2\pi \sqrt{6} \, R \, C}$$

Where:
   F is the frequency in hertz (Hz)
   R is the resistance in ohms (Ω)
   C is the capacitance in farads (F)

When the constant terms in the equation are combined, we can simplify the expression to:

$$F = \frac{1}{15.4 \, R \, C}$$

or,

$$F = \frac{0.408}{2\pi \, R \, C}$$

(Both forms are seen in popular textbooks.)

Of course, your problem in practical designs will be to find a resistor/capacitor combination that will produce a desired (and known) frequency—so this expression is backwards. Since there are fewer "standard" capacitor values, we tend to select a capacitor value, then plug in the frequency and to find the value of resistor

IF:

R1 = R2 = R3 = R

C1 = C2 = C3 = C

Then:

$$F = \frac{0.408}{2\pi RC}$$

Fig. 15-14. A)JFET RC Phase Shift Oscillator, B) op-amp version.

228

that will produce that frequency with that capacitor. In that case,
we will rearrange the equation to read:

$$R = \frac{0.408}{2 \pi R C}$$

Example

Find the resistance required to make a 1000 Hz oscillator when
a 0.01-μF capacitor is used for C1, C2, and C3 in Fig. 15-14B.

$$R = 0.408/(2 \text{ F C})$$
$$R = 0.408/(2 \times 3.14 \times 1000 \times 1 \times 10^{-8})$$
$$R = 0.408/6.28 \times 10^{-5}$$
$$R = 6497 \text{ ohms}$$

In all feedback oscillators we must ensure that the closed-loop gain
is unity or more. In the case of Fig. 15-14B, we set the gain as the
ratio of R_f/R. Analysis will show that the loss in the feedback
circuit is 1/29, so we will make the gain equal to 29 in order to
ensure oscillation. For most practical circuits, the recommendation
is $R_f = 30R$. In this case, therefore, the value of $R_f = (30)(6497)$
= 194,910 ohms.

Wein Bridge Oscillator

The Wein bridge is an RC bridge circuit similar in basic form
to the Wheatstone bridge (OK purists—sit down!). Two of the arms
of the Wein bridge are resistances, while the other two are RC
networks. One of the RC networks is a series circuit, while the other
is a parallel circuit; in both cases the R and C elements are equal
between the two networks. An example of Wein bridge oscillator
is shown in Fig. 15-15. The feedback look is degenerative (thus
stable) at all frequencies other than the oscillating frequency, which
is given by the general expression:

$$F = 1/(2 \pi \text{ R3 R4 C1 C2})$$

or, in the event R = R3 = R4 and C = C1 = C2:

$$F = 1/(2 \pi \text{ R C})$$

Wein bridge oscillators produce a sinewave output, but when R2

R1 = R2
R3 = R4 = R
C1 = C2 = C

NOTE:
I1 & R2 NOT
USED AT SAME
TIME

INCANDESCENT
LAMP

Fig. 15-15. Wein bridge oscillator.

is used the amplitude tends to be a little unstable—especially when the circuit is variable frequency. The solution is to replace resistor R2 with a small-current incandescent lamp. This lamp has a nonlinear voltage-current characteristic that tends to stabilize the output amplitude and prevents the amplifier from saturating. In most cases, I1 is a 0.040 ampere lamp, and is operated somewhat below incandescence.

Twin-Tee Oscillators

There are several other forms of sinewave oscillator that are based on RC networks. Some of these are shown in this section as twin-tee and bridged-tee oscillators. The circuit in Fig. 15-16 is an example of the twin-tee oscillator, so-called because its feedback network consists of two tee-networks. Note that these networks are of opposite types. One of them uses series resistors and a shunt capacitor, while the other uses series capacitors and a shunt resistor. The frequency is approximated by the equation below:

$$F = 1/2 \pi R C$$

Again, we can rearrange the equation to the more practical, but less familiar form:

$$R = 1/2 \pi F C$$

Example

Find the resistance needed to make a 500 Hz twin-tee oscillator when the capacitance C is 0.01-μF.

$$R = 1/(2 \times 3.14 \times F \times C)$$
$$R = 1/(2 \times 3.14 \times 500 \times 10^{-8})$$
$$R = 1/3.14 \times 10^{-5}$$
$$R = 31{,}847 \text{ ohms}$$

Two additional forms of tee-oscillator are shown in Fig. 15-17. These are two alternate forms of bridged-tee oscillator. In both cases, an RC tee-network is bridged by either a resistor or a capacitor. If the series element of a tee-network is a resistor, then the bridging element will be a capacitor, and vice-versa. As in the case of the Wein bridge oscillator shown above, the bridged-tee oscillators use a resistor and incandescent lamp to stabilize the output amplitude.

GENERATING SINEWAVES FROM SQUAREWAVE OR TRIANGLEWAVE SOURCES

Some instrument designers prefer to use either a squarewave

Fig. 15-16. Twin-Tee oscillator.

Fig. 15-17. Two forms of split tee oscillator.

or trianglewave to generate a sinewave output. Most sinewave generators are unstable with respect to amplitude, so are less than desirable. On the other hand, many squarewave oscillators operate in a saturating mode, so are inherently amplitude stable. We can do this neat trick because all non-sine waveforms are made up from a series of sine and cosine waves summed together. The squarewave and triangle, for example, contain a fundamental "sinewave" and a large number of harmonics of the fundamental waveform frequency.

If we conspire to filter out all of the harmonics, saving only the fundamental frequency, the result will be a sinewave of the fundamental frequency. The purity of the sinewave can be quite good, especially if a high order of filtering is used. In most cases, either a notch filter (which passes only the fundamental), or a series of low-pass filters are used for the conversion.

Besides amplitude stability, we gain one other advantage by using the filtering method. The other forms of RC sinewave oscillator are a little hard to make variable over a wide range of frequencies. The squarewave oscillator, on the other hand, is a lot easier in this respect. We can make a variable squarewave oscillator with only one resistor variable, and then use a low-pass filter to get rid of the harmonics. It is feasible to use a single low-pass filter per octave of frequency change.

232

RC ONE-SHOTS, SQUAREWAVE, AND TRIANGLEWAVE OSCILLATORS

In this section we investigate three different types of signal generator circuit: one-shots, squarewave generators, and triangle-wave generators. All of these circuits are based on the operational amplifier, and are quite easy to design and build.

One-Shot Circuits

The one-shot, (more formally called the "monostable multivibrator") produces one output pulse of constant duration and amplitude for each and every input trigger pulse. Such circuits are used in a variety of computer applications; for example, pulse stretching, switch/keyboard debouncing, and others. The concept of pulse stretching is a little confusing to some because the actual pulse is not stretched, but rather a new (longer duration) pulse is generated. In that application, a short duration pulse is used to trigger a longer duration pulse.

Switch debouncing is another common application for one-shots. Mechanical switches don't make and break cleanly, but rather they bounce. This phenomenon places several spikes on the line for each contact closure. While most analog circuits can absorb the extra pulses, a digital circuit cannot. We can debounce the switch contacts by using the output of the one-shot circuit as the signal to indicate contact closure, and then use the switch to trigger the one-shot. The duration of the output pulse is set long enough to allow all of the contact closure spikes to die out.

Figure 15-18A shows a one-shot stage based on the operational amplifier, while Fig. 15-18B shows the timing waveform. For a moment, let's concentrate on the timing waveform—keeping in mind the definition of the one-shot given above. The one-shot stage has only one stable output state, in this case $+V_z$ (hence the name "monostable" multivibrator). At time T1 the one-shot is triggered, so the output snaps to the unstable stage ($-V_z$). It will remain in this state for only a specified period of time, T, which is defined as (T2 – T1). Now let's return to the circuit (Fig. 15-18A).

Normally, the output of the operational amplifier is HIGH (i.e., $+V_z$), but when a short pulse is applied to the trigger input, it will snap LOW (i.e., $-V_z$). Since the output has been HIGH, capacitor C2 will have to be charged to a positive voltage. This voltage would be $+V_z$, except that diode D2 clamps the voltage to about 0.7 volts. When the output snaps LOW, however, the output volt-

Fig. 15-18. A) Op-amp monostable multivibrator, B) timing diagram.

$T_O \approx 0.707 R_4 C_2$

age is $-V_z$, so the capacitor will begin to charge toward that potential. But the voltage will not reach that value, for the noninverting input of the operational amplifier is biased to some fraction ("B") of the output voltage, as set by the resistor network R7/R8. When the capacitor voltage exceeds BV_o, the output of the operational amplifier will snap back to the HIGH, stable state.

The period immediately after the output returns to the stable state is called the "refractory period", and the circuit will not respond to further trigger inputs during this period. Unless something is done, the capacitor will have to discharge under the influence of $-V_o$, R4 and C2, and that takes a long time. We can shorten the refractory period by placing diode D3 and resistor R5 in the circuit. If R5 § § R4, then the discharge is a lot faster than the charge cycle. Diode D3 conducts only when the output voltage is positive. The duration of the output state is approximately:

$$T = R4\ C2\ Ln\left[\frac{1}{1 - B}\right]$$

Where:

    T is the period in seconds (sec)
    R4 is the resistance in ohms (Ω)
    C2 is the capacitance in farads (F)
    Ln denotes the use of the natural logarithms
    B is the feedback factor determined from the voltage
    divider equation [B = R8/(R7 + R8)]

The equation is derived from the capacitor charging equation; i.e., the time required to charge capacitor C2 from $+0.7$ volts to voltage BV_o. We can simplify the equation by setting standard values for resistors R7 and R8, therefore making B standard. For example, when the resistors are equal to each other (R7 = R8 = R), then B = R/(R + R) = 1/2 = 0.5, so equation (1) becomes:

$$T = R4\ C2\ Ln\left[\frac{1}{1 - B}\right]$$

$$T = R4\ C2\ Ln\left[\frac{1}{1 - 0.5}\right]$$

$$T = R4\ C2\ Ln\left[\frac{1}{0.5}\right]$$

$$T = R4\ C2\ Ln\ [2]$$

$$T = 0.69\ R4\ C2$$

Or, in the case where R = R7 = (2 × R8), B = R/(R + 2R) = R/3R = 0.33. In this case we obtain the following simplification:

$$T = R4\ C2\ Ln\left[\frac{1}{1 - 0.33}\right]$$

$$T = R4\ C2\ Ln\ [1.49]$$

$$T = 0.4\ R4\ C2$$

There is one other common case frequently seen where B = 0.632. Using the same method as above we find that, for B = 0.632 we have a greatly simplified expression:

$$T = R4\ C2$$

To find the resistor combination R7/R8 in which B = 0.632 we can merely solve the voltage divider equation for that value, which yields relative values for R7 and R8:

$$B = \frac{R8}{R7 + R8}$$

$$0.632 = \frac{R8}{R7 + R8}$$

Working through the tedious algebra yields the following:

For B = 0.632,

$$R7 = 0.58\ R8$$

In the most cases, we need to select a standard capacitor value and the desired output duration, and then calculate the resistance that will yield that duration. The reason for this procedure is that standard capacitor values which are commonly available are fewer in number than standard resistor values. Besides, we can always use a variable resistor to obtain specific values—and it is harder to find reasonably sized variable capacitors!

Squarewave Oscillators

An operational amplifier squarewave oscillator is shown in Fig. 15-19A, and its timing diagram is shown in Fig. 15-19B. This circuit has no stable output states, so its output snaps back and forth between HIGH and LOW conditions. The noninverting input of the operational amplifier is biased by a fraction of the output voltage, while the inverting input is biased by the voltage across capacitor

Fig. 15-19. A) Astable multivibrator, B) timing diagram.

C1 (from point "A" to ground). This voltage is determined by the output voltage, and the RC time constant of R1C1.

When this circuit is initially turned on, the capacitor contains no charge, so the inverting input is at zero potential. The output will be HIGH ($+V_o$). The capacitor will begin to charge toward $+V_o$ at the rate determined by R1C1. When it reaches the bias point of the noninverting input ($+V_f$), the output will snap LOW (i.e., $-V_o$). At this time, the charging of the capacitor reverses, and the voltage will discharge from $+V_f$ towards zero, and then charge toward $-V_f$. When it reaches this negative voltage, the output again snaps HIGH, and the capacitor begins to discharge toward zero and charge toward $+V_f$ again. This oscillating cycle continues endlessly, as long as the power is applied.

The time required for one cycle is determined by the resistances in the circuit, and the value of the capacitor. In terms of Fig. 15-19A, the period of oscillation is given in general by:

$$T = 2 \, R1 \, C1 \, Ln \left[1 + \frac{2 \, R2}{R3} \right]$$

Where:

   T1 is in seconds (sec)
   C1 is in farads (F)
   R1, R2, and R3 are in ohms (Ω)

In the event that R2 = R3 (a common selection), the equation reduces to a simpler form:

$$T = 3.2 \, R1 \, C1$$

Again, we have a situation where the equation is not as easily used as it might be, so we will rearrange it to solve for the resistance:

$$R1 = T/3.2 \, C1$$

Using this version we would specify the period (T) and a standard capacitor (C1), and then calculate the resistance required.

In the event that you want to work from the frequency of oscillation, rather than the period, then simply take the reciprocal of T:

$$F = 1/T$$

Where T is in seconds and F is in hertz.

Example

What is the frequency of a 1 millisecond (0.001-second) oscillator?

$$F = 1/T$$
$$F = 1/0.001$$
$$F = 1000 \text{ Hz}$$

Similarly, we can calculate T from knowledge of F:

Example

What is the period of a 1500 Hz oscillator?

$$T = 1/F$$
$$T = 1/1500$$
$$T = 0.00067 \text{ seconds }]$$

Now, let's work a practical design problem. Calculate the resistance needed to make a 1500 Hz oscillator (i.e., one with a period of 0.00067 seconds); see Example below.

Example

Find the resistance needed to make a squarewave oscillator similar to Fig. 15-19A operate at 1500 Hz (0.00067 second) when the capacitor (C1) is 0.001-μF.

$$R1 = T/(3.2 \text{ C1})$$
$$R1 = (0.00067)/(3.2 \times 10^{-9} \mu F)$$
$$R1 = 209,375 \text{ ohms}$$

The "duty cycle" of the circuit of Fig. 15-19A is 50 percent; in other words, the output is HIGH for the same length of time as it is LOW. In some cases we might not want a 50 percent duty cycle, but rather will need either HIGH or LOW portion to be shorter than the other part of the cycle. There are two strategies that will yield different HIGH and LOW durations (see Figs. 15-20A and 15-20B).

One alternative, shown in Fig. 15-20A, is to use a different timing resistor for HIGH and LOW portions of the cycle. In this

Fig. 15-20. Methods for changing duty cycle: A) Diode method, B) adjustable method.

case, diodes are used for switching the resistors (R_a and R_b). When the output is HIGH (+ Vz), then diode D4 is forward biased, and R_b affects the waveform timing. Similarly, when the output is LOW (– Vz), then D4 is reverse biased and D3 is forward biased so R_a controls the timing of events.

The other method is shown in Fig. 15-20B. Here we add a positive or negative biasing charge to the capacitor through resistor R_{off}. The polarity of the bias voltage across C1 is determined by the setting of potentiometer R2, and can vary from V – to zero to V +. We can monitor the output of the circuit on an oscilloscope while adjusting R2 for the desired duty factor. In most applications, the potentiometer will be a multi-turn trimpot mounted on a printed circuit board unless front panel adjustment is desired.

Triangle Wave Oscillators

A triangle wave oscillator can be made by combining the squarewave oscillator with a simple Miller integrator (see Fig. 15-21A). There are other ways of making triangle and sawtooth waveforms, but they all seem to lack something compared with this method.

A simple operational amplifier version of the standard Miller integrator circuit is shown in Fig. 15-21A. The slope of the output waveform depends upon the time constant R1C1. Figure 15-21B shows a typical waveform to expect when the time constant is long compared with the period of the applied squarewave. When the squarewave is LOW (– V), then the integrator output will climb in a nearly linear manner in a positive-going direction. The linear slope is due to the fact that the input squarewave has a constant

240

Fig. 15-21. A) Miller integrator, B) input/output waveforms.

amplitude during this time. Similarly, when the squarewave goes to the opposite sense, the output of the Miller integrator will be negative-going. This cycle continues as long as there is a symmetrical squarewave applied to the input of the integrator.

An example of a trianglewave generator based on this principle is shown in Fig. 15-22. Here we have a Miller integrator connected to a comparator circuit (A1). The frequency of oscillation will be:

$$F = \frac{R2}{4\ R1\ R3\ C1}$$

CRYSTAL CONTROLLED OSCILLATORS

The most stable frequency control is provided by the piezoelectric crystal oscillator. Some crystal elements, notably

Fig. 15-22. Triangle waveform generator.

241

quartz, possess the property of piezoelectricity. This phenomenon is not found in all materials, only certain crystalline natural minerals and manmade ceramic materials. A material is said to be piezoelectric if it generates an electrical potential when deformed. When a piezoelectric element is mechanically deformed, therefore, a voltage appears across the faces of the crystal slab.

There is also an inverse version of this phenomenon. When we apply a voltage across the faces of a piezoelectric crystal slab, the crystal will mechanically deform. We can use this phenomena to make the crystal oscillate at its resonant mechanical frequency.

If a pulse is applied to the faces of the crystal slab, then the slab will vibrate back and forth in an oscillatory manner. Losses in the slab will cause the amplitude of those mechanical oscillations to die out exponentially. An analogy is the ordinary musical tuning fork. When it is struck, it will vibrate at a certain pitch (frequency), but mechanical losses in the tuning fork will cause the oscillations to die out gradually. While the pulsed crystal slab is oscillating, however, its piezoelectric property causes it to produce an AC voltage with a frequency equal to the resonant frequency.

Figure 15-23A shows a circuit symbol for the piezoelectric crystal resonator, while Fig. 15-23B shows the actual physical form the crystal will take. The cut slab of crystal material is sandwiched between two contact electrodes. In older crystals, these electrodes are attached with spring tension, while in more modern examples they are electro-deposited directly onto the surface of the crystal slab. The crystal element is then mounted to the pins protruding through the header support structure.

Figure 15-23C shows the equivalent electrical circuit of a piezoelectric element. There is a series resistance, series inductance, and a series capacitor in the circuit. There is also a parallel capacitor in the circuit. The frequency response plot of the crystal element is shown in Fig. 15-24. This graph plots the reactance-vs-frequency of the crystal. Note that there are actually two different resonant modes for the crystal: series and parallel. You might have guessed that from the fact that there are two capacitances shown in Fig. 15-23C, one each in series and parallel with the inductance.

The series resonant frequency is that at which the inductive reactance is exactly cancelled by the series capacitance. At this point, the total reactance of the crystal is zero, and the series resistance determines crystal impedance. The impedance is thus minimum for the series resonant mode.

242

Fig. 15-23. A) Piezoelectric crystal symbol, B) construction, C) equivalent circuit.

The parallel resonant frequency of the crystal is usually 1 to 15 kHz higher than the series resonant frequency. The impedance of the crystal at this frequency is maximum (see frequency F_p in Fig. 15-24). A parallel-mode crystal will operate in the series mode if a small capacitor is connected in series with the crystal. The value of the capacitor must be equal to the specified load capacitance of the crystal.

There are two oscillatory modes for the crystal element: fundamental and overtone. The fundamental mode frequency is the natural resonant frequency of the device. That is, the mechanical frequency at which the slab oscillates if stimulated. The frequency of oscillation in the fundamental mode depends upon factors such as the crystal's mechanical dimensions, style of cut, temperature and other factors.

In the overtone mode, the crystal oscillates at a frequency that is approximately an integer multiple of the fundamental frequency. Note, however, that the overtone is not a harmonic of the fundamental. If you divide the 5th overtone frequency by 5, the

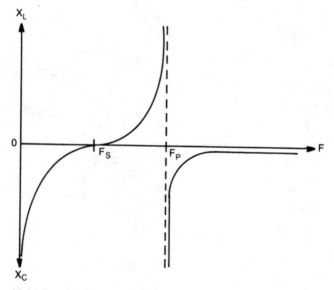

Fig. 15-24. Impedance-vs-frequency plot for a crystal.

result is not necessarily the fundamental, but a number close to the fundamental. The overtone frequencies are always approximately an odd multiple (e.g., 3, 5, 7, 9, 11), and the overtone crystal always operates in the series mode.

When ordering crystals for frequency control, we must specify not only the frequency of operation, but also the load capacitance (especially when parallel operation is contemplated).

The operating frequency of the fundamental mode is usually less than about 20 MHz; above 20 MHz the crystal slab becomes too thin for safe operation and may fracture under normal operating conditions. Fundamental frequency crystals are usually parallel mode, except below 500 kHz where it is usual to order series mode crystals. The parallel resonant conditions produces a 180 degree phase shift at the parallel resonant frequency.

The power dissipation of the crystal must be limited in order to prevent fracture damage to the slab. It is the equivalent series resistance (ESR) of the crystal that determines the power dissipation at any given level of applied signal. In a practical crystal oscillator circuit we control the dissipation by controlling the feedback signal amplitude. Most fundamental mode crystals will dissipate up to about 200 microwatts (μW), although it is usually considered good design practice to make the actual dissipation somewhat less than this maximum.

244

Low frequency crystals, i.e., those operating below 1000 kHz, usually have a dissipation rating of 100 μW. It is generally regarded as good practice to limit the dissipation to 50 μW in order to improve the frequency stability of the crystal oscillator. It is not good practice to obtain huge amounts of rf power from the crystal oscillator. It is necessary to limit the dissipation of the oscillator circuit, and then use an amplifier to build up the levels.

Examples of Crystal Oscillator Circuits

The piezoelectric crystal resonator is essentially equivalent to a complex LC circuit (with series and parallel sections), so can be used in many of the same circuits as ordinary LC resonant tank circuits. In many such cases, the sole difference between the circuits is that the crystal replaces the LC tank circuit.

Figure 15-25 shows a simple circuit for low frequency crystals, specifically the 100 kHz "marker" frequency crystal shown. The oscillator operates at a fundamental frequency of 100 kHz, but its harmonics go high enough to make the circuit useful for marker service in communications receivers and shop alignment.

The active element in this circuit is a bipolar NPN transistor, such as the popular 2N2222 (the actual type is not as critical as

Fig. 15-25. Crystal oscillator.

one might assume). Almost any transistor with a beta in the 50 to 125 range, and a gain-bandwidth product of 50 MHz or more will work. The reason for specifying the 2N2222 is that they are easily available in hobbiest and amateur radio parts stores.

Note that the circuit configuration for the 100 kHz oscillator is similar to the familiar LC Colpitts oscillator discussed earlier. Feedback level is a function of the capacitive voltage divider, C1/C2. The ratio of these two capacitors is a trade-off between output amplitude, stability, and the dissipation of the crystal element. The collector load is provided by resistor R1, and the transistor is biased in the collector-base manner by resistor R2.

The crystal element Y1 shunts the feedback voltage divider, and is in series with variable capacitor C3. This capacitor is used to set the actual operating frequency of the calibrator. Almost any crystal operating frequency can be "pulled" a little bit with a series or parallel capacitor.

The output signal is coupled to the load through a low-value capacitor, C4. This circuit should be lightly loaded in order to minimize changes in frequency due to load variations. When this is a critical concern, the solution to follow the oscillator with a buffer amplifier stage.

There are several ways to calibrate the operating frequency of the oscillator. If you have a digital frequency counter, then use it. Once upon a time, this would have been a ridiculous statement, for digital frequency counters were a multi-kilobuck investment. But today, there are a number of low cost DFC's and DFC kits on the market that all but surpass the performance of instruments that were top dollar only a decade or so ago.

The alternate way to calibrate the circuit is to use a radio station. If you lack a communications receiver, then use an AM broadcast band radio. The radio should be tuned to a radio station that has a frequency divisable by 100. In my area, for example, I could use WTOP on 1500 kHz. If you have a communications receiver, then use WWV in Fort Collins, Colorado or or WWVH in Hawaii. These stations transmit accurate frequency checks on 5, 10, 15, and 20 MHz. Select the highest of these frequencies that produces a strong local signal. Since the error is multiplied along with harmonics, it is best to use the 15 or 20 MHz frequency, even if you have to time your work to when "skip" is in. Adjust C3 for zero beat with the radio station. If you have an "S-meter" on the receiver, then zero beat is indicated as the meter pointer wobble gets slower and slower—which means the two frequencies are closing in on one another.

The classic Miller oscillator used a crystal resonator in the grid of a vacuum tube, and a parallel tuned LC tank circuit in the plate. In Fig. 15-26 we see the updated version using a junction field effect transistor (JFET) in place of the tube.

The crystal resonator is connected between gate and common in Fig. 15-26, and is in parallel with a 10 Megohm resistor. Bias for the JFET is set by source resistor R2, which is bypassed for RC to make a low impedance to ground for signal. In most cases, the reactance of the capacitor is set at less than 1/10 the resistance it bypasses.

The drain terminal of the JFET in this circuit is connected to the resonant tank circuit. The actual resonant frequency of this circuit only approximates the crystal frequency. The oscillation of this circuit takes on different properties depending upon which side of the crystal frequency the tank is set to. In most cases, the tank circuit is tuned to the low side of the resonant point.

On first glance the Miller oscillator doesn't look like an oscillator at all because there is no apparent feedback network. This circuit depends upon "hidden" capacitances in the circuit, shown in dotted lines in Fig. 15-26 as C_{gs} and C_{dg}. These capacitors rep-

Fig. 15-26. Miller oscillator.

Fig. 15-27. Variable crystal oscillator.

resent the interelectrode or interelement capacitances of the active device, in this case JFET Q1. The two capacitances are the gate-source and drain-gate capacitances, respectively. They form a capacitive voltage divider that feeds back some of the output signal to the input. In some cases, one or more of these capacitors are augmented with small value (i.e., several picofarads) external capacitors.

Another example of a simple Colpitts crystal oscillator is shown in Fig. 15-17. This circuit is somewhat different from the other Colpitts configuration, but may be actually more popular. This circuit is designed to operate over the 1 to 20 MHz frequency range, although my personal experience is that operation is a little "iffy" above 15 MHz or so.

Feedback is controlled by the C2/C3 capacitor voltage divider, the fact that identifies this as a "Colpitts" oscillator. As in the previous Colpitts example, almost any NPN transistor can be used. In fact, a PNP is useful if the DC power supply polarity is reversed.

Finally, in Fig. 15-28 we see an example of an overtone oscillator. Recall that an overtone oscillator will operate on an odd multiple of the fundamental frequency (or near multiple). An

248

exception in some circuits is that the 4th overtone, which is an even number. The case for the overtone crystal is marked with the intended overtone frequency, not the fundamental frequency.

The crystal element is connected between the base terminal on the transistor and ground. Capacitor C1 is used to improve the feedback on transistors that have insufficient internal capacitance. In some cases, C1 will not be needed, but where it is used it must be mounted as close to the body of the transistor as possible.

The tank circuit (L1/C3) in the collector circuit of the transistor is tuned approximately to the desired overtone frequency. The emitter resistor bypass capacitor is designed to have a reactance of approximately 90 ohms at the frequency of operation; values for common frequencies are given below.

Frequency (MHz):	50	150	220	450
Capacitance (pF):	36	12	8	3

TTL CLOCK CIRCUITS

In this section we discuss digital TTL electronic "clocks". These circuits are not time-of-day clocks, but rather are used to

Fig. 15-28. Overtone oscillator.

create pulse trains that time or synchronize digital electronic circuits. In almost all cases the digital clock will produce either a squarewave or trapezoidal wave. In this article we deal with digital clock circuits based on *transistor transistor logic* (TTL) integrated circuits.

TTL Basics

Let's review the basics of TTL digital electronics. The TTL logic family was probably the first really successful family of integrated circuit digital devices. Previous families (e.g., RTL and DTL) never really achieved the widespread popularity enjoyed by TTL devices. As in any standard logic family, the TTL devices uses standardized input and output circuits and standardized logic levels.

Digital electronics is binary in nature; that is, it permits only two possible states. These states represent the digits of the binary (base-2) number system, 1 and 0. These two states are sometimes called "HIGH" and "LOW", a convention that will be followed in this article. Figure 15-29 shows the logic levels for TTL devices. The HIGH level (which represents "1") is attained when the input or output voltage is greater than +2.4 volts but less than +5 volts. The LOW (or "0") condition is represented by a grounded state, which means a potential between 0 and 0.8 volts.

Some literature will tell you that TTL devices will operate at supply and signal potentials greater than +5 volts DC, but I strongly recommend against it. Such potentials are in what I call the "Groan Zone", and will drastically shorten the life of TTL devices.

Negative potentials are another area of concern because negative potentials will kill the device. Although it is possible for power supply potentials to be applied backwards, the most likely case for oscillator builders is inappropriately connected capacitors and other components that store energy. Negative potentials will destroy the TTL device, so must be avoided—and it is possible to avoid the problems simply by being cautious in physical construction.

Figure 15-30 shows the standardized input and output circuits of TTL devices. Any logic family works together because inputs and outputs can be interconnected with only a conductor—no impedance matching or other considerations are necessary. The TTL input acts as a 1.6 milliampere current source, while the output operates as a current sink. When we connect a TTL output to a TTL input, therefore, the circuits match perfectly.

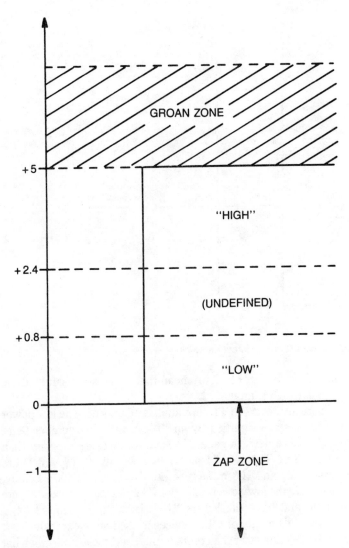

Fig. 15-29. TTL operating zones.

Interfacing TTL devices becomes a matter of making sure that drive requirements are met. The requirements of a single standard input (1.6 mA @ +2.4 to +5.0 volts) is the interfacing unit; the single input is said to have a "fan-in" of 1, which represents the requirements of a single TTL input. The output will have a fan-out number which defines the number of standard fan-in-1 TTL inputs that it can drive. Thus, when we say that the standard TTL

Fig. 15-30. TTL input/output interfacing.

output has a fan-out of 10 we mean that it can successfully drive 10 standard TTL inputs.

There are several TTL sub-families of devices. The low power TTL devices are signified by an "L" in the type number (e.g., 74L00). These devices generally have lower drive capacity than regular TTL. There is also the high speed TTL, which are designated by an "H" in the type number (e.g., 74H00). Finally, we also have the low-power Schottky TTL devices. These are quite common, and have the letters "LS" is the type number (e.g., 74LS00). This family of TTL devices is used quite extensively. It has Schottky diodes at the input circuits, and these are somewhat sensitive to static electricity. Therefore, it is recommended that you handle most LS-series TTL devices almost as gingerly as you would handle CMOS devices.

Figure 15-31 shows a method for converting a standard transistorized oscillator circuit to TTL. The oscillator is a standard Colpitts oscillator built around an NPN transistor (Q1). The feedback level is set by the usual capacitor voltage divider, C1 and C2. The oscillator frequency is set by Y1, a piezoelectric crystal. Some small degree of frequency control is permitted by varying capacitor C3. Crystals oscillate on a frequency that is determined in part by the capacitance they operate into, so capacitor C3 will "pull" the operating frequency of the crystal.

The output stage is an LM-311 comparator. This type of IC is basically a differential amplifier with too much gain. In any differential amplifier, the output voltage is a function of the difference between the two input voltages. Thus, when the voltages on pins 2 and 3 are equal, the difference is zero so the output voltage will be zero also. But when these two voltages are different by only a couple of millivolts, the output voltage will be nonzero. Since the gain could be 10,000 to 100,000, the output will saturate anytime the differential input voltage is nonzero. In Fig. 15-31, the pin 3 input is grounded so sees a zero potential. Thus, when the signal applied to pin no. 3 crosses zero or is negative, the output voltage will be zero. But when the input voltage is positive, then the LM-311 output voltage is HIGH.

The LM-311 device has what is called an "open collector" output stage. This means that it requires a pull-up resistor to V +. Although the LM-311 is able to operate with output potentials greater than +5 VDC, we use +5 VDC here to make the circuit compatible with TTL circuits. For higher potentials simply scale the resistor upwards in value proportional to the increase in voltage—up to about +15 VDC.

Another way to do this same trick is to use a TTL chip called a "Schmitt trigger". The circuit is shown in Fig. 15-32A, while the

Fig. 15-31. TTL-izing a transistor oscillator.

Fig. 15-32. Using a Schmitt trigger A) circuit, B) waveform.

operation of the Schmitt trigger is shown in Fig. 15-32B. The operation of the Schmitt trigger follows this simple rule: the output will snap HIGH when a positive-going input signal crosses a certain threshold, and will snap LOW when the input signal crosses a lower threshold in a negative-going direction. Thus, when a signal from a transistor oscillator crosses +1.7 volts in the positive-going direction the Schmitt trigger output snaps HIGH, and it will drop LOW again when the input drops to +0.9 volts again. In both cases, Figs. 15-31 and 15-32, the AC sinewave from a transistor oscillator or other signal source will produce a train of squarewave at the output of the Schmitt trigger.

A small collection of TTL oscillators is shown in Fig. 15-33. These are all based on TTL inverters or inverter-connected NAND or NOR gates (note: a NOR or NAND gate will become an inverter if all inputs are connected together). The first circuit, Fig. 15-33A,

Fig. 15-33. Several TTL clock circuits.

255

is RC-timed. That is, the oscillating frequency is determined by capacitor C1 and resistors R1 and R2. In this particular configuration the resistance is adjustable in order to continuously vary the operating frequency over a small range. If only a single fixed frequency is needed, then replace potentiometer R1 and resistor R2 with a single fixed resistor.

A disadvantage of any RC timed oscillator is that the operating frequency is neither stable or accurate. These problems are partially overcome using piezoelectric crystals, as in Figs. 15-33B and 15-33C. The circuit in Fig. 15-33B is based on the 7400 TTL NAND gate. The 7400 IC contains four independent NAND gates, of which two are used for the oscillator and a third operates as a buffer stage; all three NAND gates are connected as inverters. The two gates (A & B) used in the oscillator circuit are each biased with a 1.5 kohm resistor, and are isolated from each other by coupling capacitor C2. The operating frequency is set by crystal Y1, and may be varied slightly with capacitor C1.

The version shown in Fig. 15-33C is similar, and is based on TTL inverters. These stages are from a package of six called a "hex inverter" (e.g., 7404). Again, one stage is used as an output buffer, while the oscillating stages are self-biased.

Special TTL Oscillators

There are several TTL oscillators available on the market, but some of them are a little hard for hobbyists and amateurs to obtain. In Fig. 15-34 we see a circuit based on the MC-4024P voltage controlled oscillator. This Motorola circuit is not to be confused with the CMOS device called the "4024", which is a binary counter in another logic family. The MC-4024P device contains two TTL voltage controlled oscillator stages. In this circuit we show only one stage being used.

There are two ways to set the overall operating frequency (other than controlling the input voltage): capacitor or crystal. In most cases, the capacitor is used and will have a value of approximately (300/Frequency) picofarads. For frequencies above 2000 kHz, however, we may also elect to use a crystal to set the frequency (both methods are shown in Fig. 15-34).

The circuit in Fig. 15-34 has the control voltage input (pin no. 2) connected to +5 volts for fixed frequency operation. If we want to make the frequency variable, then we would use a circuit such as Fig. 15-35. In this case, we have a potentiometer and fixed resistor connected in a voltage divider to set the control input volt-

256

Fig. 15-34. TTL MC4024P crystal oscillator.

age at some potential less than +5 volts—with a subsequent pulling
of the operating frequency. When a capacitor is used in place of
crystal Y1, the frequency can be pulled over a 3:1 range. The control
input is grounded for AC signals by capacitor C1, which also serves
to stabilize the voltage (hence the frequency).

Fig. 15-35. TTL MC4024P crystal oscillator VCO.

257

CMOS CLOCK CIRCUITS

The last type of oscillator circuit that we will consider is the CMOS digital logic oscillator. Like their TTL counterparts, the CMOS oscillator is often used as a "clock" in digital electronic circuits for timing and synchronization. There are some significant differences between TTL and CMOS, however.

Perhaps the biggest difference, and one that drives the other differences, is that CMOS devices are made from metal oxide field effect transistors (MOSFETs) instead of the NPN/PNP bipolar transistors used in TTL devices. An implication of this fact is that CMOS devices draw considerably less current than TTL devices (microamperes instead of milliamperes), although generally at the cost of operating speed.

Figure 15-36 shows the basic CMOS inverter circuit used to make most of our oscillator circuits. The basic inverter of Fig. 15-36A is representative of the older "A-series" CMOS devices, and consists of a pair of MOSFET transistors. Device Q1 is an N-channel MOSFET, while Q2 is a P-channel MOSFET. These devices operate in opposite "polarity" from one another: a HIGH applied to the common (parallel-connected) gate input will cause one of these transistors to saturate while the other turns off.

The newer B-series CMOS device is shown in Fig. 15-36B. There is a resistor in series with a gate, and at least one diode shunting the gate. The reason for these components is electrostatic discharge protection. Normal A-series devices have a gate breakdown voltage of less than 100-volts. Since static built up on clothing and tools can easily reach several kilovolts, you can damage A-series CMOS by touching the device. The B-series provides some protection in this matter.

Most CMOS devices have part numbers on the 4000-series. RCA devices are denoted as "CD-4xxx", while Motorola uses "MC-144xxx"; other manufacturers use the "4xxx" designations. The A-series devices will either have an "A" suffix on the part number, or no suffix at all. For example, "4001" and "4001A" both denote an A-series device. On the other hand, the "4001B" is the same CMOS device in B-series, and is preferred for most applications.

There are other differences between A-series and B-series CMOS devices. The B-series offers faster rise and fall times for squarewave pulses, and will drive larger loads than most A-series counterparts. Although there may be some applications where the

Fig. 15-36. CMOS inverter A) A-series, B) B-series, C) equivalent circuit.

characteristics of the A-series provides some advantage, it is generally true that the B-series device is preferred.

The operation of the CMOS inverter can be seen in Fig. 15-36C. The transistors in the CMOS device can be modeled as gated resistors. In Fig. 15-36C the channel resistance of Q1 is represented by R1, while the channel resistance of Q2 is represented by R2. The output is at the junction of the two resistances. CMOS devices operate from two power supplies: V + and V − , both of which can be from roughly 4 volts to 18 volts. It is common to find V + set to + 5 volts and V − set to 0-volts (i.e., ground); this method makes the device usable in circuits where TTL power supplies are already available. We could also select − / + 5 volts, − / + 12 volts, or some non-symmetrical potentials (e.g., − 5 volts and + 12 volts). Note that the operating parameters of most CMOS devices are measured at − / + 10 volts, and operation away from these potentials is somewhat different.

Let's take a look at what happens when a signal is applied to the input in Fig. 15-36C. Assume that V + is 5 volts (which also represents HIGH), and V − is grounded (which represents LOW). When the input is LOW, then the output is HIGH. This means that resistor R1 has a low value (less than 2 kohms), while R2 is very high (1 megohm). In this case, the output terminal is effectively connected to V +, so the output is HIGH. Alternatively, when the input is HIGH, the resistances reverse: R1 becomes very high and R2 becomes very low, so the output terminal is effectively connected to ground.

The reason for the very low current requirement is that the V − to V + path is always a very high resistance in series with a very low resistance. The only time when both resistances are moderately low is when the output is in transition (between LOW and HIGH, or HIGH to LOW).

Our first CMOS clock circuit is shown in Fig. 15-37; the circuit is in Fig. 15-37A while the timing diagram is shown in Fig. 15-37B. The 4093 device is a Schmitt trigger, and four such triggers come in the 4093 package. A Schmitt trigger is a special device that operates only on specific input voltages. Assuming that the power supply is +5 volts and ground, the inputs operate on positive-going signals that pass +2.9 volts, and on negative-going signals that drop below +2.3 volts. The device must see a positive-going input prior to the negative-going threshold crossing, or it won't work.

The device shown in Fig. 15-37A is a two-input NAND gate, so we can either tie the two inputs together, or use one of the inputs as a control element. When the switch is open, the input is at V +, so the device will operate normally; when the switch is closed the input is grounded, so the device will sit there dormant with the output at HIGH. For our description of operation, assume that the switch is open.

At turn-on, capacitor C1 is discharged (0-volts) so the input (point "A") is at ground potential, or LOW. The output is HIGH at this instant (see time T_0 in Fig. 15-37B). The voltage across the capacitor begins rising as C1 charges from the output potential (through R1). The voltage (V_{C1} in Fig. 15-37B) will eventually reach the positive-going threshold, V1. At this time, the input sees a HIGH, so will cause the output to drop LOW. This change of condition removes the charge source from R1C1, so C1 will begin to discharge through R1 and the output terminal of the 4093. When the voltage decreases to V2, the output again snaps HIGH, and the cycle repeats itself.

A second CMOS oscillator, this one a little more common, is shown in Fig. 15-38. This circuit can be made using a CMOS inverter, or NAND and NOR gates connected as inverters (i.e., with both inputs tied together). The operating frequency of this circuit is set by the resistor-capacitor combination according to the approximate formula: $F = 1/(2.2 \times R1 \times C1)$. The resistor (R1)

Fig. 15-37. CMOS astable circuit based on 4093: A) circuit, B) timing diagram.

261

Fig. 15-38. CMOS RC-timed astable circuit.

can be anything from 3.3 kohm to several megohms, and the capacitor (C1) from roughly the values shown up to about 10 μF.

The output waveform of this device is a little unsymmetrical, especially when B-series devices are used. The problem with the B-series device is that the internal diode protection disturbs the situation, so some means must be provided to isolate the R1C1 combination from the protected input. Figures 15-39A and 15-39B show how this is done: resistor R2 (in both cases) isolates the RC network from the diode-protected input, and cures the asymmetry

Fig. 15-39. Improved versions of the CMOS astable circuit.

problem in the waveform. These circuits are generally preferred over that of Fig. 15-38.

A pair of voltage controlled oscillators are shown in Figs. 15-40A and 15-40B. These circuits produce an output frequency that is partially dependent upon the value of the control voltage applied to the input. In Fig. 15-40A this voltage is applied in a manner to bias the input of the inverter (and the junction of the RC network) to some static value. The charge/discharge cycle that sets the operating frequency takes longer or shorter duration depending upon the applied voltage. The Vin = 0 set frequency is found according to the formula above.

The circuit in Fig. 15-40B works a little differently. In this case, a MOSFET transistor shunts the RC junction to ground, thereby affecting the charge/discharge timing. The control voltage is applied to the gate input of the MOSFET transistor.

Fig. 15-40. Two CMOS voltage controlled oscillators.

The 4046 device shown in Fig. 15-41 is an IC CMOS voltage controlled oscillator (VCO). The chip is listed in the data sheet as a phase-locked loop, but it can be operated as a VCO. The control voltage input to pin no. 9 is provided by a potentiometer or other source. The operating frequency is determined by C1 and R1, and to some extent by the optional resistor R3. The value of R1 can be anything from 3,000 to 1,000,000 ohms, while the capacitor may be from 100 pF to 10 μF. It is possible to make an oscillator that operates over a range of 1000:1 with the 4046 device.

The use of R3 varies the minimum and maximum operating frequencies. Ordinarily, with R3 open-circuited, the operating frequency will vary from DC to the maximum set by R1/C1 (to 1 MHz or so). If R3 is used, however, the operating frequency range

Fig. 15-41. CMOS PLL oscillator.

Fig. 15-42. Crystal controlled CMOS oscillator.

changes: there will be a minimum and a maximum different from the above rule.

Our final example is the crystal oscillator shown in Fig. 15-42. This circuit uses two sections of a 4001 NOR gate, both of which are inverter connected (but by different means). One section of the 4001 operates as the oscillator, while the other is a buffer amplifier. The crystal oscillator is an exception to the digital types of circuit shown elsewhere in this chapter. The reason is that the 4001 is biased into the linear region by R1, so acts as an amplifier (gains of 10,000 are possible with buffered B-series devices). The operating frequency is set by the crystal (Y1) operated in its parallel resonant mode. The RC components are selected to provide a 180 degree phase shift at the crystal's parallel resonant frequency. This circuit will provide good stability and accuracy of output frequency.

Electrostatic Damage

There is one problem with CMOS devices: they are damaged by static electricity. Almost everyone is familiar with static shocks.

If you walk across a rug, or if you wear certain kinds of cloth garments, then a static charge builds up on your body. Touch something that is grounded, and—zzzappp!!—there will be a small but startling spark. That spark represents very little electrical current, but the voltages can easily reach several kilovolts. These potentials are sufficient to blow CMOS devices.

If you are lucky, then the damage from the electrostatic discharge (ESD) is immediate, and can be repaired. But it is often the case that the damage takes awhile to develop, and will not show up until later. ESD is the source of many unexplained premature failures of equipment that contains a lot of CMOS devices.

The way to avoid ESD damage to CMOS devices is elimination of the static electricity. The usual method is to keep everything at the same potential. CMOS devices come packaged in either a special static-free plastic tube, or on a small piece of carbonized foam rubber. The idea is to keep all the IC pins at the same voltage. If you don't take the device out of its protected environment until time to use, and then make sure that you discharge body static through several megohms of resistance to ground, then there is little possibility for damage.

Chapter 16

Microcomputers: A Basic Primer

The field of digital electronics has been with us for a long time, and many of its techniques grew out of vacuum tube concepts and electromechanical relay switching circuits. The field didn't really take off, however, until the invention of the integrated circuit digital logic element. For several years the digital electronics field expanded rapidly, and finally in the mid-1970s, the microprocessor chip became available. From the microprocessor we were able to build microcomputers—small desktop (and smaller) units that mimicked the capability of whole rackfulls of equipment. Microcomputers have literally revolutionized the electronics industry and several related industries. Once, knowledge of computers was limited to a few people in the backwaters of engineering. But today, it is a requirement that all types of technological and scientific personnel must be smart about the microcomputer and its applications. Nowhere is this trend more evident than in electronic engineering, and especially in the fields of instrumentation and control. The advantage of the microprocessor or microcomputer is that it allows us to "computerize" our designs at low cost. So what, you say? Well, consider the advantages of computerization. We gain a greater degree of control over factors such as drift when we use the digital implementation of a design. In addition, if we universalize our design sufficiently, then it is possible to gain immense reconfigurability in the digital implementation. Sometimes, we find that relatively major redesigns can be accomplished with only a little hardware and some software patches. On older designs using analog methods, even a minor change in the design configuration could have profound impact on the physical hardware: new printed circuit boards had to be laid out, new circuitry tested, and so forth. With the microcomputer implementation, however, we are sometimes able to use only minor hardware changes along with software changes to accomplish big changes of mission.

A side benefit of computerization of no small significance to many companies is the fact that logistics supply of computerized systems if often easier. The reason for this is that a single computer printed circuit board will work for a wide variety of products. A technical person need only plug in the right Read Only Memory (ROM) or read the correct program from a disk or other long-term storage media and the machine changes its identity.

The microprocessor is the heart of the microcomputer and it is a *large-scale integrated circuit* (LSI) (typically in a package containing 40-pins). The microcomputer IC contains all of the circuitry needed to make the central processing unit (CPU) of a programmable digital computer (in addition, some chips also provide limited internal memory and I/O capability). Most microprocessors, especially the early designs, required at least a few other external chips in order to make a complete digital computer. The 8048, 8051 and 8052 by Intel, however, are examples of "single-chip computers" that contains all of the microprocessor CPU functions found in any μP chip, plus a limited amount of internal random access read/write memory (RAM), 2K of read only memory, and a pair of input/output (I/O) ports. If your application requires no capability other than the internal memory and I/O, then the 8048 is capable of providing your entire computer needs.

The purpose of this book is to introduce you to digital electronics in detail and computers in general. Some of the computers discussed in these pages are sold complete with cabinet and keyboard, while others are "bare board" devices that are "stripped" for use as a component in some other system. Both types of devices are properly called microcomputers. In fact, the problem is where to draw the line.

MICROPROCESSORS: A SHORT HISTORY

Intel Corporation is usually credited with the invention of the microcomputer in 1972. The story is usually told that Intel was contracted to design and produce a four-bit programmable "controller" that would allow the user to control traffic signal lights. That unit supposedly became the now obsolete (but still seen occasionally) 4004 four-bit microprocessor device. Oddly enough, the original contractor (so the industry legend goes) did not want the device, so Intel was free to go on the open market with it . . . to the benefit of the rest of us. The next unit was the 8008 device, which was basically an eight-bit 4004. Later, the now-famous 8080A device was designed and offered to the market. It was the 8080A that became the microprocessor of choice for many thousands of engineers, technicians, hobbyists, and others who were eager to break into the field. It was amazing how the extremely costly 8080A device plummeted in cost until now they are all but giving it away; that's called the "silicon learning curve" theory of pricing.

After the 8080A became a huge success, another company got onto the bandwagon: Zilog, Inc. began to produce the all but legendary Z80 microprocessor. The Z80 was designed using the same general philosophy as the 8080A, and in fact, was able to use the entire 8080A repertoire of machine code instructions—plus a few dozen additional instructions that the 8080A did not recognize.

The Motorola 6800 was probably the next device on the market, and it used a different architectural philosophy than either the 8080A or the Z80 devices. The MOS technology family of 65xx devices (of which the 6502 is the most famous but not only example) was next on the scene, and became immensely popular. The Apple *II* is based on the 6502 device. Currently, almost every semiconductor company makes at least one microprocessor either of their own or under a "second source" license from a primary vendor.

SOME DEFINITIONS

The terminology of microcomputer/microprocessors is used by almost everyone today, but even so it bears a little repeating since there are so many who tend to use the words incorrectly.

Programmable Digital Computer

A programmable digital computer contains a "memory" section that is used to store both data and instructions. The central processing unit (CPU) in a programmable digital computer responds to a certain set of binary numbers that it recognizes as "commands" or "instructions." These binary numbers are stored in memory. The job that the computer will perform is determined by these instructions and both the manner and order of their use in the "program." Since the programmer can insert these instructions in memory, the computer is said to be *programmable*. It is "digital" because it works with binary devices (such as electronic switches) that can implement binary (i.e., base 2) arithmetic. The alternative to programmable computers is non-programmable or fixed-program computers, while the alternative to digital computers is analog computers (an all but obsolete form). These latter forms were once used extensively, and consisted of amplifiers, integrators/differentiators (based on operational amplifiers) and a lot of other analog electronic circuitry.

Bit

The word "bit" is an acronym made up from letters of the words "**Bi**nary digi**ts**." A "bit" is a digit of the binary, or base 2, numbers system: either 0 or 1. These digits are represented by discrete voltages or currents in the electronic circuitry. For example, in the popular TTL logic system of digital devices (74xx/74xxx devices), the zero is represented by a voltage from 0 to 0.8 volts, while the one is represented by a voltage of +2.4 to +5 volts.

BYTE

A "word" consisting of eight bits; e.g., 11010011 is an example of an eight-bit, or "one byte," binary number.

Microprocessor

A microprocessor is an integrated circuit that contains all of the necessary components to form the central processing unit of a programmable digital computer. The microprocessor is not to be confused (as it often is) with the microcomputer. The latter term is used to denote a fully programmable digital computer that uses a microprocessor as the CPU. Thus, a microcomputer will contain a microprocessor plus whatever additional circuitry is needed to make it qualified to operate as a full computer.

Microcomputer

A microcomputer is the smallest element that is truly

identifiable as a full-fledged programmable digital computer. It is, by definition, built around a microprocessor CPU rather than other chips. It will contain, in addition to the microprocessor chip, other chips to fill out the functions of the computer. For small computers, these extra chips might be only two or three LSI special-purpose devices that contain a small amount of RAM, perhaps some ROM and, of course, the I/O functions. On larger machines, the additional chips may number 50 to 100, or even more. Just what comprises a microcomputer varies somewhat. For example, the small $4.5 \times 6.5 \times 6.5$ inch O.E.M. printed circuit board STD-Bus computer is a microcomputer, as are the Heath H-8 Apple *II* and IBM-PC. Yet, at the same time, we see huge rack-mounted computers that look more like minicomputers also designated as "microcomputers." The reason? They are based on a microprocessor CPU chip.

Minicomputer

The minicomputer has traditionally been regarded as a scaled-down mainframe data processing computer. Examples of true minicomputers include the now-obsolete (but once ubiquitous) Digital Equipment Corporation PDP-8 and their nearly legendary PDP-11 family of computers. These machines have not in the past used microprocessor chips in the CPU, but rather discrete logic elements, mostly TTL. It is not uncommon for a minicomputer to require 30 to 40 amperes of current from the +5 volt regulated DC power supply.

In the past, minicomputers have been more powerful than microcomputers. They would, for example, sport bit lengths of 12 to 32 bits rather than 4 or 8 bits found in most microcomputers. In addition, they would operate at clock speeds of 8 to 12 MHz rather than 1 to 10 MHz used in the microcomputer. Today, however, the distinctions have faded, and are now almost indistinct. We now see microcomputers that are housed in rack-mounted cabinets six-feet high, and minicomputers that are desktop units. The minicomputer may also use an LSI CPU chip that looks suspiciously like a microprocessor, albeit of special manufacture. The DEC PDP-11 format was implemented in the LSI-11 microprocessor chip and wound up in such computers as the Heath Hill. It is now quite difficult to draw the line of distinction between the minicomputer and the microcomputer.

Mainframe Computers

The big horse used in the data processing department of large companies and most banks is called a mainframe computer. The IBM 370 is an example of a large mainframe computer. Mainframe computers are derisively called "dinosaurs" by the ignorant elitists among the microcomputer fraternity. But be well aware that this dinosaur is not about to become extinct like its reptillian namesake. If the mainframe computer is a "dinosaur," then the microcomputer is a mere lizard!

Interfacing

"Interfacing" is probably among the most abused words in the English language. Salesmen have been known to write to me stating that they would like to come down and " . . . interface with" me. What's that? Are they going to insert an RS-232 connector in my mouth and plug in? I usually discard such letters immediately; any document that comes across my desk with "interface" or its derivatives used in that manner will get a "C – " for starters.

Interfacing (when the word is correctly used in a computer equipment context), is the art of taking a microcomputer and wedding it with other circuits and equipment in order to solve some particular problem. At least one author has defined interfacing as the art of replacing hardware logic with software, and that is a reasonable explanation in my opinion. It is interesting to note that interfacing is a relatively simple process involving understandable and reasonable rules and procedures; it doesn't ordinarily require loads of "smarts." Yet, it is often in the art of interfacing that engineers, technicians and other luminaries make their reputation among the non-engineering personnel. In a university research setting, for example, one can make large strides up the respect ladder by being able to work the black magic that makes a computer control an experiment or improve an instrument.

SELECTING THE RIGHT MICROPROCESSOR

The gut-feeling selection of a microprocessor will usually lead us to the chip with the greatest capability, the longest data bus word length and the most superior instruction set. But that might also be the least economical choice! The selection of a microprocessor chip should be made with regard to the application at hand (seems reasonable, but for many it is a revolutionary idea!), and not on an emotional level at which the desire for the "best" supersedes everything else. If a brand new 16-bit microprocessor chip seems exciting, then take a second look at the situation before becoming committed. If, for example, an 8-bit microprocessor will do the same job, and still cover all planned or anticipated expansion, then why go to the trouble and expense of a 16-bit machine? But on the other hand, if the application and its planned growth strains the 8-bit machine, then it might be cheaper in the long run to go ahead and spend the money up front on the 16-bit device.

Proper regard must be given to the potential for expansion of the system. If you lock yourself into a machine that does not operate fast enough, or have a good enough instruction set, or cannot support sufficient memory, then you will be hard pressed to make the machine deliver additional capability when additional features must be called for. Also, consider the matter of memory size. It turns out that one never has enough memory. Sometimes this problem is due to the natural tendency for programmers to use up the entire memory of the computer. In fact, one computer designer claims that it is wise to tell the programmers about only half of the memory that is really available! Otherwise, they will use up all of what you allocate despite your protestations that they should leave room for expansion. If it turns out that the programmer genuinely requires additional memory, then it is possible to "add" memory without too much bloodletting.

THE TAB AUTOMATIC COMPUTER

This section was added to the book on the advice of a friend of mine who noted that, all too often, microcomputer books assume a familiarity with computers in general that is simply not there in all readers. Thus, we have a need for a section that discusses the programmable digital computer as a generic device. The "machine" of this section is a universal computer, representative of the entire set of programmable digital computers, and not any one manufacturer's offerings. It is a hypothetical machine, and not available anywhere. It, like an androgynous being, has elements of all kinds within its body. We call this hypothetical computer the TAB Automatic Computer, or "TABAC" for short.

Figure 16-1 shows the block diagram of TABAC. We will first describe each of the blocks in this diagram, and then progress to a discussion of how TABAC processes a typical computer program.

Like any programmable digital computer, TABAC has three main parts: *central processing unit* (CPU), *memory*, and *input/output* (I/O). There are other functions found in certain specific machines, but many of these are either special applications of these main

Fig. 16-1. Computer block diagram (TABAC).

sections, or are too unique to be considered in a discussion of a "universal" machine.

The central processing unit (CPU) controls the operation of the entire machine. It consists of several necessary subsections which are described in greater detail below.

Memory can be viewed as an array of "pigeonholes," or "cubbyholes" (Fig. 16-2) such as those used by postal workers to sort mail. Each pigeon hole represents a specific address on the letter carrier's route. An address in the array can be uniquely specified (identifying only *one* location by designating its *row* and *column*. If we want to specify the memory location (i.e., pigeonhole) at row 3 and column 2, then we would create a *row-3, column-2* address number, which in this case is "32."

Each pigeonhole represents a unique location in which to store mail. In the computer, the memory location stores not pieces of paper, but a single binary "word" of information. In an eight-bit microcomputer, for example, each memory location will store a single eight-bit binary word (e.g., 11010101). The different types of memory device are not too terribly important to us here, except

Fig. 16-2. Memory illustration.

in the most general terms. To the CPU and our present description of how it operates, it doesn't matter much whether memory is *random access read/write memory* (RAM), *read only memory* (ROM).

There are three main lines of communications between the memory and the CPU: *address bus, data bus*, and *control logic signals*. These avenues of communications control the interaction between memory and I/O, on the one hand, and the CPU on the other, regardless of whether the operation is a read or write function. The address bus consists of parallel data lines, one for each bit of the binary word that is used to specify the address location. In most eight-bit microcomputers, for example, the address bus contains sixteen-bits. A 16-bit address bus can uniquely address up to 2^{16}, or 65,536 different eight-bit memory locations. This size is specified as "64K" in computers, not "65K" as one might expect. It seems that lower case "k" is used almost universally in science and engineering to represent the metric prefix *kilo*, which represents the multiplier 1000. Thus, when someone tells you that their computer has "64K" of memory, you might expect it to contain 64×1000, or 64,000 electronic pigeonholes. But you would be wrong in that assumption. Long ago computerists noted that 2^{10} was 1,024, so they determined that their "k" would be 1,024, not 1,000! This means that a 64K computer will contain a total of 64 $\times$ 1024, or 65,535 electronic pigeonholes in which to stuff data. To differentiate "big-k" (1024) from "little-k" (1000), standard computerist shorthand uses upper case "K" rather than lower case "k."

Be aware that the size of memory which can be addressed doubles for every bit added to the address bus. Hence, adding one bit to our sixteen-bit address bus creates a seventeen-bit address bus that is capable of addressing up to 128K of memory. Some "8-bit" machines which have 16-bit address buses are made to look larger by tactics which create a pseudo-address bus. In those machines, several 64K memory banks are used to simulate continuously addressable 128K, 256K or more machines.

The data bus is the communications channel over which data travels between the main register (called the *accumulator* or *A-register* in the CPU and the memory. The data bus also carries data to and from the various input and/or output ports. If the CPU wants to read the data stored in a particular memory location, then that data is passed from the memory location to the accumator (in the CPU) by way of the data bus. Memory write operations are exactly the opposite: data from the accumulator is passed over the data bus to a particular memory location.

The last memory signal is the *control logic* or *timing* signal. These signals tell memory if it is being addressed, and whether the CPU is requesting a *read* or *write* operation. The details of the control logic signals vary considerable from one microprocessor chip to another, so there is little that we can say at this point. Later in this book you will be introduced to both the 6502 and Z80 standard signals (which are representative of two different architectures), but for others I recommend either the manufacturer's literature, or my own book titles the *8-bit and 16-bit Microprocessor Cookbook*, TAB Books catalog number 1643.

The input/output (I/O) section of the TABAC computer is the means by which the CPU communicates with the outside world. An input port will bring data in from the outside world and then pass it over the internal data bus to the CPU, where it is stored in the accumulator. An output port is exactly the opposite: it passes accumulator data from the data bus to the outside world. In most cases, the "outside" world consists of either peripherals (e.g., printers, video monitors), or communications devices (e.g., MODEMs).

In some machines, there are separate I/O instructions that are distinct from the memory-oriented instructions. For example, in the Z80 machine there are separate instructions for "write to memory" and "write to output port" operations. In Z80-based computers, the lower eight bits (A0 - A7) of the address bus are used during I/O operations to carry the unique address of the I/O port being called (the accumulator data will pass over both the data bus and the upper eight bits—AB to A15—of the address bus). Since there are eight bits in the unique I/O address, we can use up to 2^8, or 256 different I/O ports that are numbered from 000 to 255. In the 6502-based computer, there are no distinct I/O instructions. In those machines (e.g., Apple *II*), the I/O components are treated as memory locations. This technique is called *memory-mapped I/O*. Input and output operations then become memory-read or memory-write operations, respectively.

The CPU is literally the heart and soul of TABAC. Although there are some differences between specific machines, all will have at least the features shown in our TABAC computer diagram of Fig. 16-1. The principal subsections of the TABAC computer are: *accumulator* or *A-register, Arithmetic Logic Unit* (ALU), *Program*

Counter (PC), *Instruction Register* (IR), *Status Register* (SR) or *Processor Status Register* (PSR), and *Control Logic Section*.

The accumulator is the main register in the CPU. With a few exceptions, all of the instructions use the accumulator as either the source/destination of data or the object of some action (e.g., unless otherwise indicated, an ADD instruction always performs a binary addition operation between some specified data and the data in the accumulator).

Although there are other registers in the CPU (the Z80 is loaded with them!), the accumulator is the main register. The main purpose of the accumulator is the temporary storage of data being operated on or transferred within the machine. Note that data transfers to and from the accumulator are non-destructive. In other words, it is a misnomer to tag such operations as "transfers," because in actuality they are "copying" operations. Suppose, for example, the program calls for us to "transfer" the hexadecimal number $8F from the accumulator to memory location $A008. After the proper instruction ("STA $A008" in 6592 assembly language) is executed, the hex number $8F will be found in *both* the accumulator *and* memory location $A008.

It is important to remember that accumulator data changes with every new instruction! If we have some critical datum stored in the accumulator, it is critical that we write it to some location in memory for more permanent storage. This function is sometimes performed on the "external stack" or in some portion of memory set aside by the programmer as a "pseudo-stack."

The arithmetic logic unit (ALU) contains the circuitry needed to perform the arithmetic and logical operations. In most computers, the arithmetic operations consist of addition and possibly subtraction, while the logical operations consist of AND, OR, and Exclusive-OR (XOR).Note that even subtraction is not always found! In some computers, there is no hardware arithmetic function other than addition. The subtraction function is performed in software using two's complement arithmetic. Multiplication and division are treated as multiple additions or subtraction unless the designer has thought to provide a hardware multiply/divide capability.

The program counter (PC) contains the address of the next instruction to be executed. The secret to the success of any programmable digital computer such as TABAC is its ability to fetch and execute instructions *sequentially*. Normally, the PC will increment appropriately (1, 2, 3, or 4) while executing each instruction (i.e., 1 for a one-byte instruction, 2 for a two-byte instruction, etc.). For example, the "LDA,N" instruction on the 6502 microprocessor loads the accumulator with the number "N." In a program listing, we will find the number "N" stored in the next sequential memory location from the code for the LDA portion (called the operations code, or "op-code" for short):

```
0205    LDA
0206    N
```

At the beginning of this operation, the PC contains "0205," but after execution the PC will contain the number "0206" because LDA,N is a two-byte instruction.

There are several ways to modify the contents of the PC. One way is to let the program execute sequentially: the PC contents will increment for each instruction. We can also activate the *reset* line, which forces the PC to either location $0000, or some other specific location (often at the other end of the memory, for example $FFFA in the 6502). Another method is to execute either a JUMP or a conditional-JUMP instruction. In this latter case, the PC will contain the address of the "jumped-to" location after the instruction is executed. Finally, some computers have a special instruction that will load the PC with a programmer-selected number. This "direct-entry" method is not available on all microprocessors, however.

The instruction register is the temporary storage location for the instruction codes that were stored in memory. When the instruction is fetched from memory by the CPU, it will reside in the instruction register until the next instruction is fetched.

The instruction decoder is a logic circuit that reads the instruction register contents, and then carries out the intended operation.

The control logic section is responsible for the "housekeeping" chores in the CPU. It issues and/or responds to control signals between the CPU and the rest of the universe. Examples of typical controls signals are: memory requests, I/O requests (in non-memory-mapped machines), read/write signalling, interrupts and so forth.

The status register, whose bits are also sometimes referred to individually as the "status flags," is used to indicate to the program and sometimes the outside world the exact status of the CPU at any given instant. Each bit of the status register represents a different function. Different microprocessor chips use different sets of status flags, but all will have at least a *carry flag* (C-flag), and a *zero flag* (Z-flag). The C-flag indicates when an arithmetic or logical operation results in a "carry" from the most significant bit of the accumulator (B7), while the Z-flag indicates when the result of the present operation was zero (00000000); typically, Z = 1 when result is zero, and C = 1 when a carry occurred.

We have now finished our tour of the CPU of the TABAC computer. This discussion in general terms also describes most typical microprocessors. Although various manufacturers use different names for the different sections, and some will add sections, almost all microprocessors are essentially the same inside of the CPU—at least in respect to basics.

Operation of TABAC

A programmable digital computer such as TABAC operates by sequentially fetching, decoding, and then executing instructions stored in memory. These instructions are stored in the form of binary numbers. In some early machines, there were two memory banks, one for program instructions and one for data. The modern computer, however, uses a single memory bank for both instructions and data.

How, one might legitimately ask, does dumb ol' TABAC know whether any particular binary word fetched from memory is data, an instruction or a binary representation of an alphanumeric character (e.g., ASCII)? The answer to this instruction is key to the operation of TABAC: cycles!

TABAC operates in *cycles*. A computer will have at least two discrete cycles: instruction fetch and execution (in some machines the process is more sophisticated, and cycles are added). While the details differ from one machine to another, the general operation is similar for all of them.

Instructions are stored in memory as binary numbers ("op-codes"). During the instruction fetch cycle, an op-code is retrieved from the memory location specified by the program counter, and stuffed into the instruction register inside the CPU. The CPU assumes that the programmer was smart enough to arrange things according to the rules, so that the datum fetched from location ABCD during some instruction cycle in the future is, in fact, an instruction op-code. It is the responsibility of the programmer to arrange things in a manner so as to not confuse the poor, dumb TABAC.

During the first cycle, an instruction is fetched and stored in the instruction register (IR). During the second (i.e., next) cycle, the instruction decoder circuit inside of the CPU will read the IR, and then carry out the indicated operation: this is called the "execution cycle," while the first cycle was the "instruction fetch/decode" cycle. The CPU then enters the next instruction fetch cycle, and the process is repeated. This process is repeated over and over as long as TABAC is operating.

Each step is synchronized by an internal "clock" that is designed to make things remain orderly.

From the above description, you might be able to glimpse a truth concerning what a computer can or cannot do. The CPU can shift data around, perform logical operations (AND, OR, XOR), add two N-bit numbers, and sometimes—but not always—subtract, all in accordance with a limited repertoire of instructions encoded in the form of binary words. Operations in TABAC (and all other microcomputers) are performed *sequentially* through a series of discrete steps. The secret to whether or not any particular problem is suitable for computer solution depends entirely on whether or not a plan of action (called an "algorithm") can be written that will lead to a solution through sequentially executed steps. Most practical instrumentations, control, measurement, or data processing problems can be so structured—a fact which accounts for the dramatic rise of the microcomputer in those fields.

COMPUTERS IN INSTRUMENTATION
AND CONTROL APPLICATIONS

It has become something of a truism that the microcomputer is a superior means for implementing control and instrumentation circuits. What makes the microprocessor/microcomputer superior to other methods? After all, isn't the microprocessor just another integrated circuit? Well, I suppose you could say that is true if you are also willing to claim that a 500-lb Bengal Tiger is "just another cat!" The microcomputer is a full-fledged programmable digital computer. In some cases, the chip inside is a microprocessor (e.g., Z80 or 6502), while in others the chip is a "single-chip computer" (e.g., 8048). Microprocessors become microcomputers by adding external memory and I/O circuitry.

While many readers are familiar with the business of solving instrumentation/control problems using analog circuits, or ordinary digital logic circuits and devices such as relays and switches, we find that the microprocessor is a little different. There are two main goals in microprocessor-based instrumentation:

1. Replacement of digital logic devices with software, and
2. Replacement of some (as many as possible) analog circuits with software.

Before we hear too many howls of anguished protest over the second point, let me hasten to point out—and dispense with—a myth about digital instrumentation that is sometimes held by analog circuit designers. It is often claimed that "digital" contains a built-in error, i.e., the so-called *discrete quantization error*. This error is, indeed, inherent in the digital implementation of any circuit function. It is absolutely true that digitization forces us to inherit a certain error rate. But the argument that analog circuitry is therefore somehow "more accurate" because it permits an infinite number of discrete values between limits is not supportable. Analog circuits also contain substantial errors: amplifier gain error, offset voltage error, temperature drift, aging, and readout or display resolution. The digital implementation of a circuit function often produces a superior accuracy because it eliminates the analog errors while keeping the digitization errors low and indentifiable. Incidentally, the digitization error is reduced by a factor of two for every bit that is added to the data word length.

Another myth is the notion that, by using a computer, all of your problems are solved. It is a common fallacy that computers solve problems: that is a lie—computers do not solve problems. The computer can no more "solve a problem" than a saw or hammer can "build a house." In designing electronic instrumentation using microprocessor/microcomputer techniques you must first solve the problem at least at the flow chart level. Only after you have produced a step-by-step plan of action (i.e., created an algorithm) that will do the job can you begin to select the components that will be used. Just as the analog circuit designer begins with a block diagram solution of his problem on paper, the microcomputer-oriented designer must begin with an algorithm on paper, and sometimes a system block diagram.

Chapter 17

Typical Microprocessors

The age of the microprocessor chip began in 1972 when Intel introduced the 8008 device. The 8008 is crude by today's standards and was almost immediately supplanted by another Intel offering, the 8080. The newer device was to become one of the most popular microprocessor chips on the market for a long time. The Z80 which is manufactured by Zilog, Inc., uses all of the 8080 instructions and adds a few of its own. But the Z80 is not merely an updated version of somebody else's toy; it is a completely new device in its own right. Do not think that the Z80 is the be-all and end-all of all microprocessors, however. The reason that this device was selected as one of our examples in this elementary text is that the author is familiar with the Z80, as well as the 6502, our other selection.

THE Z80

The internal structure of the Z80 (Fig. 17-1) is not much unlike the block diagram of the last chapter. Indeed, many computers could use similar block diagrams. The Z80 device uses an eight-bit data bus and a sixteen-bit address bus. This configuration means that the Z80 will address up to 2^{16}, or 65536, different memory locations. The Z80 has 18 special-purpose or general-purpose eight-bit registers, and four 16-bit special-purpose registers. The main register set is grouped into two groups: the main register set and the alternate register set.

Fig. 17-1. Z80 register arrangement.

SPECIAL-PURPOSE REGISTERS

The special-purpose registers are the program counter (PC), the stack pointer (SP), index registers IY and IX, interrupt vector (I), and memory refresh (R).

Program Counter

As in all computers, the PC holds the 16-bit address of the instruction currently being fetched from a memory location. The program counter contents are normally either incremented by one count or is loaded with the memory address called for by a jump instruction.

Stack Pointer

A stack is a section of external memory organized as a *last-in-first-out* (LIFO) file. Data can be pushed onto the stack or popped off the stack. The SP register contains the 16-bit address of the *top* location of the external memory stack.

Index Registers

Indexed addressing uses the contents of an index register and a byte stored in memory one location away from the op-code for an index addressing instruction in order to calculate the address of a memory location. The Z80 contains two index registers: IX and IY.

Interrupt Address Register (I)

The Z80 uses an address mode in which the effective location of an instruction is given by the combination of two bytes that form a two-byte, or 16-bit, word. One byte (high order) is supplied by the contents of the I register, while the other is supplied by the interrupting device.

Memory Refresh Register (R)

A certain type of external memory device, called dynamic memory, requires periodic refreshing. The R register contains data to refresh the memory.

GENERAL-PURPOSE REGISTERS

The Z80 contains two accumulators, two flag registers, and six general-purpose registers. The accumulator, of course, is the main register. The flag register is used to contain the bits that indicate the status of the CPU, or the result of the operations.

The general-purpose registers are labeled B, C, D, E, H, and L in the main register set, and B', C', D', E', H', and L'. The two accumulators are labeled A and A', while the two flag registers are F and F'. Figure 17-1 summarizes the Z80 register set.

Figure 17-2 shows the Z80 pinouts. The Z80 is packaged in a 40-pin DIP IC package. The pin functions are summarized in Table 17-1.

All digital programmable computers work their magic because computers are synchronous devices. Such a device operates in step with a clock signal, which in a digital circuit is a chain of pulses. For example, in the Z80 microprocessor, which uses TTL compatible inputs and outputs, the clock pin sees a train of pulses that alternate from LOW (0 to 0.8 volts) to HIGH (over +2.4 volts) at a frequency of no more than 2.5 MHz for Z80, 4 MHz for Z80A, and 6-8 MHz for certain other variants of the Z80 design.

The programmable digital computer operates in cycles (see the TABAC section in Chapter 16), and those cycles are defined by the clock. Figure 17-3 shows the instruction fetch cycle for the Z80 microprocessor. Each of the pulses are clock pulses, and serve to

Fig. 17-2. Z80 pinouts and definitions.

Table 17-1. Z80 Pin Functions.

A0-A15 (Address Bus)	Tri-state output, active high. A0 - A15 constitute a 16-bit address bus. The address bus provides the address for memory (up to 64K bytes) data exchanges and for I/O device data exchanges. I/O addressing uses the 8 lower address bits to allow the user to directly select up to 256 input or 256 output ports. A0 is the least significant address bit. During refresh time, the lower 7 bits contain a valid refresh address.
D0-D7 (Data Bus)	Tri-state input/output, active high. D0-D7 constitute an 8-bit directional data bus. The data bus is used for data exchanges with memory and I/O devices.
$\overline{M_1}$ (Machine Cycle one)	Output, active low. $\overline{M_1}$ indicates that the current machine cycle is the OP code fetch cycle of an instruction execution. Note that during execution of 2-byte op-codes, $\overline{M_1}$ is generated as each op code byte is fetched. These two byte op-codes always begin with CBH, DDH, EDH or FDH. $\overline{M_1}$ also occurs with $\overline{IORQ}$ to indicate an interrupt acknowledge cycle.
$\overline{MREQ}$ (Memory Request)	Tri-state output, active low. The memory request signal indicates that the address bus holds a valid address for a memory read or memory write operation.
$\overline{IORQ}$ (Input/Output Request)	Tri-state output, active low. The $\overline{IORQ}$ signal indicates that the lower half of the address bus holds a valid I/O address for a I/O read or write operation. An $\overline{IORQ}$ signal is also generated with

an $\overline{M1}$ signal when an interrupt is being acknowledged to indicate that an interrupt response vector can be placed on the data bus. Interrupt Acknowledge operations occur during $\overline{M1}$ time while I/O operations never occur during $\overline{M1}$ time.

$\overline{RD}$
(Memory Read)

Tri-state output, active low. $\overline{RD}$ indicates that the CPU wants to read data from memory or an I/O device. The addressed I/O device or memory should use this signal to gate data onto the CPU data bus.

$\overline{WR}$
(Memory Write)

Tri-state output, active low. $\overline{WR}$ indicates that the CPU data bus holds valid data to be stored in the addressed memory or I/O device.

$\overline{RFSH}$
(Refresh)

Output, active low. $\overline{RFSH}$ indicates that the lower 7 bits of the address bus contain a refresh address for dynamic memories and the current $\overline{MREQ}$ signal should be used to do a refresh read to all dynamic memories.

$\overline{Halt}$
(Halt state)

Output, active low. $\overline{HALT}$ indicates that the CPU has executed a $\overline{HALF}$ software instruction and is awaiting either a non maskable or a maskable interrupt (with the mask enabled) before operation can resume. While halted, the CPU executes NOP's to maintain memory refresh activity.

$\overline{WAIT}$
(Wait)

Input, active low. $\overline{WAIT}$ indicates to the Z-80 CPU that the addressed memory or I/O devices are not ready for a data transfer. The CPU continues to enter wait states for as long as this signal is active. This signal allows memory or I/O devices of any speed to be synchronized to the CPU.

$\overline{INT}$
(Interrupt Request)

Input, active low. The Interrupt Request signal is generated by I/O devices. A request will be honored at the end of the current instruction if the internal software controlled interrupt enable flip-flop (IFF) is enabled and if the $\overline{BUSRQ}$ signal is not active. When the CPU accepts the interrupt, an acknowledge signal ($\overline{IORQ}$ during $M1$ time) is sent out at the beginning of the next instruction cycle. The CPU can respond to an interrupt in three different modes that are described in detail in section 5.4 (CPU Control Instructions).

$\overline{NMI}$
(Non Maskable Interrupt)

Input, negative edge triggered. The non maskable interrupt request line has a higher priority than $\overline{INT}$ and is always recognized at the end of the current instruction, independent of the status of the interrupt enable flip-flop. $\overline{NMI}$ automatically forces the Z-80 CPU to restart to location 0066H. The program counter is automatically saved in the external stack so that the user can return to the program that was interrupted. Note that continuous $\overline{WAIT}$ cycles can prevent the current instruction from ending, and that a $\overline{BUSRQ}$ will override a $\overline{NMI}$.

$\overline{RESET}$

Input, active low. $\overline{RESET}$ forces the program counter to zero and initializes the CPU. The CPU initialization includes:

1(Disable the interrupt enable flip-flop
2) Set Register I = 00H
3) Set Register R = 00H
4) Set Interrupt Mode 0

During reset time, the address bus and data bus go to a high impedance state and all control output signals go to the inactive state.

$\overline{BUSRQ}$
(Bus Request)

Input, active low. The bus request signal is used to request the CPU address bus, data bus and tri-state output control signals

287

	to go to a high impedance state so that other devices can control these buses. When BUSRQ is activated, the CPU will set these buses to a high impedance state as soon as the current CPU machine cycle is terminated.
BUSAK (Bus Acknowledge)	Output, active low. Bus acknowledge is used to indicate to the requesting device that the CPU address bus, data bus and tri-state control bus signals have been set to their high impedance state and the external device can now control these signals.
φ	Single phase TTL level clock which requires only a 330 ohm pull-up resistor to +5 volts to meet all clock requirements.

synchronize the operation of the machine. The period of each pulse is called a "T-cycle." The instruction is broken into three sub-cycles labelled M1, M2, and M3. The M1 cycle occupies four clock periods or T-cycles, while M2 and M3 each occupy three T-cycles. The use of the three cycles is as follows:

> M1—Op-Code Fetch
> M2—Memory Read
> M3—Memory Write

The M1 cycle is also called the machine cycle. A pin on the Z80 microprocessor (designated the M1 pin) drops LOW whenever the computer is in the machine cycle, i.e., when it is fetching the op-code for the next instruction from memory.

The M1 cycle is a busy period, as shown by Fig. 17-4. During the first half of the M1 cycle, that is during T1 and T2, the Z80 is in the op-code fetch mode. Since these op-codes are stored in

Fig. 17-3. Z80 timing cycle.

memory, the op-code fetch is essentially a memory read operation. Thus, during times T1-T2 both the MREQ (memory request) and RD (read) lines are LOW in addition to the M1 line. Also during T1-T2, the Program Counter (PC) contents are passed along the address bus (A0 - A15) to designate the memory location of the op-code being fetched. The data (i.e., an op-code in this case) from that location is passed over the data bus to the Z80.

The second half of the M1 cycle consists of T-cycles T3-T4. This period is used to operate the refresh system for dynamic memory. Such memory devices decay rapidly unless refreshed, so the Z80 provides the ability to refresh such memory. Devices such as the 4116, 4132, 4164 and 41256 are 16K, 32K, 64K and 256K memory chips, respectively, and are used extensively in modern microcomputers. During the T3-T4 period of the M1 cycle, the refresh address is output to the address bus; the M1 signal goes HIGH (i.e., inactive) and RFSH line drops LOW (active). The MREQ signal also goes active for a portion of T3-T4.

The input/output cycles are shown in Fig. 17-4. On the Z80, the inputs are treated as read operations, while the outputs are treated as write operations (which they are). The RD signal indicates reads by dropping LOW, while the WR indicates write operations also by dropping LOW. The IORQ signal will go LOW simultaneously with RD or WR (depending on the operation). During the equivalent memory operations, the MREQ line is LOW instead of IORQ.

Fig. 17-4. Z80 I/O cycle.

The Z80 microprocessor allows up to 256 I/O ports (numbered 0 to 255). During both input read and output write cycles the eight bit port address is placed on the lower order byte of the address bus (A0 - A7). Also active (i.e., LOW) during these operations is the IORQ line. During the read cycle (an input operation) both RD and IORQ are LOW. Data applied to the input port is gated onto the data bus (and hence into the Z80 accumulator) during the last portion of the third T-cycle. Input data must remain stable for at least this period.

During the output write cycle (also shown in Fig. 17-4), IORQ and WR are simultaneously active. The output data from the Z80 accumulator is placed on the data bus immediately following T2, and remains stable for the entire period T1-T3. The control signals (IORQ and WR) do not become active until the beginning of T2, so stable data is assured.

Memory read and write operations are similar to I/O operations, except that all sixteen bits of the address bus (A0 - A15) are used, so a total of 65,536 (i.e., 64K) of memory can be accessed. Also, the MREQ control signal replaces the IORQ control signal. Both memory read and write cycles require three T-cycles (T1-T3).

During the memory read cycle, the MREQ and RD control signals drop LOW, and the sixteen bit address appears on A0 - A15. The read cycle takes three clock cycles, and the data memory will appear on the bus during the last part of T3.

During the write cycle, the MREQ and WR control signals become active (LOW). The data from the accumulator is placed on the data bus at the end of T1, and will remain stable for T2 and T3. Thus, the data being written to memory will be stable during the period when WR is active. Like the read operation, the memory element being addressed is selected by a sixteen bit code on A0 - A15.

Z80 INSTRUCTION SET

A measure of microprocessor's power is the nature and extent of the machine code instruction set. Compared to the original 8080A, which was the predecessor of the Z80, the Z80 has an almost awesome instruction (which is less awesome in light of later 16-bit machines). The instructions for the Z80 microprocessor are grouped into the following sets:

1. Load and Exchange
2. Block Transfer/Search

3. Arithmetic and Logical
4. Rotate and Shift
5. Bit Manipulation
6. JUMP, CALL and RETURN
7. Input/Output
8. CPU Control

According to the official Zilog count, there are 158 distinct Z80 instructions, almost twice the 8080A count. But when you break apart the grouped instructions according to bit or register operated on, then the 158 expands to well over 400.

Load and Exchange Instructions. These instructions are used for data movement both within the CPU and between the CPU and the rest of the world. Amongst the load instructions are two basic forms. One type involves purely internal Z80 data moves, i.e., between registers. The other type involves data moves between the Z80 CPU and external memory. The load instructions are further subdivided into 8-bit and 16-bit groups. Although the Z80 is an 8-bit machine, there are instructions that permit the transfer of two bytes of data into and out of Z80 register pairs. For the purpose of these instructions, internal Z80 registers are grouped AF, BC, DE and HL.

The exchange instructions are used to swap the contents between analogous register pairs. For example, "EX DE,HL" swaps the contents of register pairs DE and HL. These instructions include the following:

    EX AF, AF'
    EX DE, HL
    EX (SP), HL
    EX (SP), IX
    EX (SP), IY
    EXX

The last instruction, EXX, swaps the entire main bank with the prime bank. In other words, each register swaps its contents with the contents of the analogous prime register (e.g., the contents of register A' go to A, and the contents of A go to A').

Block Transfer and Search Instructions. This group of instructions permits the computer to either transfer (in one operation) an entire block of data, or to search a block for specific data. The sixteen bit register pairs are used to designate the starting locations of the origin and destination addresses, as well as a byte counter (which specifies the number of bytes transferred). Members of this group are LDI, LDIR, LDD, LDDR (all transfer instructions), API, CPIR, CPD and CPDR (all search instructions). For all of these instructions the sixteen bit register pairs are used as follows:

HL Source Address
DE Destination Address
BC Byte Counter

The block transfer instructions will transfer an N-byte long block of data stored in sequential locations. The definitions of LDI, LDIR, LDD and LDDR follow:

LDI (Load and Increment). Transfers one byte of data from the memory location specified by (HL) to the locations specified by (DE). This instruction will then increment both DE and HL, while decrementing BC.

LDIR (Load, Decrement and Repeat). This instruction is the same as LDI, except that it repeats until finally BC is decrementing to 00H. By loading the initial source and destination addresses in HL and DE, respectively, and the number of bytes to be transferred into BC, an entire block of data can be transferred.

LDD and LDDR. Identical to the LDI and LDIR, except that the HL and DE are decremented instead of being incremented.

The block search instructions compare the contents of the accumulator with the contents of a memory location specified by (HL). The result of the comparison is indicated by the flag register (F). The following are block search instructions: CPI, CPIR, CPD and CPDR.

CPI (Compare and Increment). Single operation comparison instruction. Following SPI, (HL) is incremented and (BC) is decremented.

CPIR (Compare, Increment and Repeat). This instruction repeats the CPI operation until one of two conditions exists: 1) the contents of BC are 00H, or 2) the contents of the memory location pointed to by (HL) exactly matches the data in the accumulator; (HL) is incremented.

CPD/CPDR. Same as CPI and CPIR, except that the contents of HL are decremented instead of incremented.

Arithmetic and Logical Instructions. This group of instructions performs the computing functions of the computer.

There are two groups of arithmetic and logical instructions: eight-bit and sixteen-bit. In the 8-bit arithmetic group are the addition (ADD), add with carry (ADC), subtract (SUB), and subtract with carry (SBC). The 8-bit logical group consists of AND, OR, or XOR instructions. Also included in this 8-bit classification are compare (CP), increment (INC) and decrement (DEC) instructions.

The same basic set of ADD, ADC, SBC, INC and DEC instructions are used in the 16-bit arithmetic/logical group, but instead of the 8-bit registers (A, B, C, D, D, H and L), and single memory locations, they use the register pairs (HL, IX, IY, BC, DE and SP).

There is also a series of instructions that decimal adjust the accumulator data (DAA). These instructions can be used to allow multi-precision BCD numbers, signed or unsigned binary numbers or two's complement signed numbers.

There are four additional instructions in the arithmetic/logical group: complement accumulator (CPL), negate accumulator (NEG), complement carry flag (CCF), and set carry flag (SCF). The CPL instruction causes the number in the accumulator to be complemented. This means that the 1's become 0's and the 0's become 1's. The NEG instruction causes the contents of the accumulator to be expressed in two's complement form. The CCF instruction causes the carry flag to be complemented. If the carry flag is 1, then it will become 0, and if it is 0 it will become 1. The SCF instruction will cause the carry flag to be set (i.e., forced to 1).

Rotate and Shift Instructions. The rotate and shift group instructions include RLC, RRC, RL, RR, SLA, SRA, SRL, RLD, RRD, RLCA, RRCA, RLA and RRA. These instructions move specified bits left or right.

Bit Manipulation Instructions. One of the things that makes the Z80 one of the better early microprocessor designs is the existence of the bit manipulation instructions. We can test a bit 1 or 0, we can reset a bit (RES) to 0, or we can set (SET) a bit to 1. The particular bit tested, set or reset can be in any specified register (A, B, C, D, E, H and L), or any memory location. In the latter case, we may use either indexed or register indirect addressing of the selected memory location. There are quite a large number of individual instructions in this group, because we can select any of eight bits (0 - 7), seven different registers, or memory locations specified by HL, IX and IY register pairs.

JUMP, CALL, and RETURN Instructions. A digital computer executes instructions in a sequential manner. In the ordinary course of events, the Program Counter is incremented one to several counts every time an instruction is executed. The number of counts incremented is determined by the number of bytes required for the particular type of instruction. Although this sequential execution is one of the powerful aspects of digital computers, it would also limit the range of possible problems that could be solved to those amenable to sequential processing (although parallel processing is now being developed). It would be impossible to perform most operations requiring even the simplest decision. Even the simple matter of inputting data would be impossible. In those operations an input port is connected to a keyboard. We create a loop, using a JUMP instruction that tests the strobe bit (usually bit B7), and if none is found jumps back to the beginning of the loop. If a strobe is found, on the other hand, the program is allowed to fall through to the next instruction (usually an input instruction in the keyboard case). But this is merely a trivial example. Most problems requiring decision logic on the part of the computer could not be performed without the use of the JUMP, CALL and RETURN instructions. The instructions in this group include JP, JR, CALL, DJNZ, RET and RETN.

A JUMP (mnemonic: JP) instruction is a branch to a subroutine at some address other than the next address in the sequence. The address at which the next instruction (i.e., the first instruction of the subroutine) is to be found is loaded into the Program Counter (PC). We may use either of three addressing modes: immediate extended, register indirect, and relative.

Each of the different types of JUMP instructions are keyed to certain conditions that are reflected by the status bits of the flag (F) register. The conditions that may be specified by the selection of the op-code include carry, noncarry, zero, nonzero, parity even, parity odd, sign negative, sign positive and unconditional.

The conditional JUMP instructions look for the status of the appropriate bit of the flag register. If the condition is met, then the JUMP operation occurs.

In immediate extended addressing, the JUMP occurs to a 16-bit memory address specified by the two bytes following the JUMP instruction. If the condition called for is met, then the program control will be shifted to the location specified by the following two bytes.

Register indirect addressing allows us to store the 16-bit address of the first instruction in the subroutine we wish to execute in one of the three double registers. Either the HL, IX or IY register pairs can be specified by appropriate selection of the op-code. The register indirect JUMP instructions are all unconditional.

A relative addressing JUMP (JR mnemonic) instruction exists for each of the following conditions: carry, noncarry, zero, nonzero, and unconditional. In this type of JUMP instruction the next instruction for the program to execute (i.e., the first instruction op-code in the subroutine) is specified by the current contents of the PC added to a displacement integer, e. The value of e can be anything in the range -126 to $+129$, as measured from the address of the instruction op code (rather than the location of the displacement integer).

In the case of one of the unconditional JUMP instructions (all forms of JUMP used in the Z80 will recognize an unconditional "condition"), the Program Counter is loaded immediately with the two bytes immediately following the op-code for the jump. The second byte of these three-byte instructions becomes the low order byte of the PC address, while the third byte of the instruction becomes the high order byte of the PC address. Since the contents of the PC are now changed, program control is transferred to the instruction located at a memory location specified by the Program Counter.

There is also one special form of JUMP instruction that is very useful: DJNZ. This mnemonic stands for "decrement register B and jump if it is nonzero." This instruction allows us to use the B register as a byte counter. We load register B with an integer equal to the number of times that we wish to execute a subroutine. The program control will transfer to the subroutine specified by a displacement integer e (this is relative addressing) as soon as the jump occurs again. This will continue until B counts down to zero. If B was preloaded with zero, then the program will jump and loop through all 256 integers before terminating when zero is again encountered.

An example of a possible application of this type of instruction is signal averaging. If we want to average 100 points of data, for example, we can next an input instruction inside of a DJNZ loop. The B register is loaded with 100, and is decremented with each execution. When all 100 data points are input, then the B register decrements one more time to zero, terminating the operation. The program then falls through to the next instruction in the sequence. Note that the relative displacement integer e is expressed in the form of a two's complement number.

The CALL instruction is a special case of the JUMP instruction.

If the CALL is used, then the address of the memory location immediately following the CALL instruction is loaded into an external memory stack (pointed to by the SP register contents). This allows us to branch to a subroutine, and then return to the main program sequence.

The return (RET) instruction is a reverse CALL instruction, and is used to return to the main program once the subroutine jumped to by the CALL instruction is finished. The RET instruction is usually the last instruction in the subroutine. When this instruction is encountered, the Program Counter is loaded with the contents of the external memory stack (again pointed to by the SP register). This will be the address of the first instruction that instigated the subroutine branch.

There are two specialized RETURN instructions, RETI and RETN. These are for returning to main program control after the servicing of an interrupt and nonmaskable interrupt, respectively.

Input/Output Instructions. The I/O instructions are the immediate input and immediate output instructions (IN A,n and OUT A,n). In these instructions, the operand n is the eight-bit address of one of 256 (i.e., 000 - 255) possible ports. This address will appear on the lower eight bits (A0 - A7) of the address bus, while the contents of the accumulator appear on the upper eight bits of the address bus (A8 - A15). The input or output data are passed over the eight bit data bus to, or from, the accumulator (respectively).

In the case of register indirect addressing, the contents of the C-register specify the eight-bit address of the selected port. This address (i.e., the contents of C) are passed over the lower byte of the address bus to signal the device being selected. Also at this time, the contents of the B register are passed over the high order byte of the address bus. This is analogous to the immediate I/O instructions, except that the sources of the data passed to the address bus are different.

The block input instructions include INI, INIR, IND and INDR. The block instructions include OUTI, OTIR, OUTD and OTDR. They are analogous to the memory block transfer instructions, except that they use the contents of register pair HL to point to an eight-bit I/O address at an external memory location. In these instructions, register B is used as a byte counter. As in the register indirect case above, the contents of the C register hold the address of the I/O port. Note that this means that the contents of register pair HL point to a location in external memory. The contents of this location are then loaded into register C. When the actual transfer takes place, the contents of C are then placed on the low order byte of the two-byte address bus. Also, as in other register indirect I/O instructions, the contents of the B-register (i.e., byte counter) are placed on the high order byte of the address bus.

CPU Control Instructions. There are seven instructions in the Z80 set that are used exclusively for the control of the CPU: NOP, HALT, DI, EI, IM0, IM1 and IM2. The NOP instruction is a "no operation" instruction. During the execution of a NOP instruction, the CPU will allow the programmer to set any of three interrupt modes. The zero mode (IM0) causes the Z80 to think that it is an 8080A microprocessor. The IM1 mode causes program control to transfer automatically to location 0038H when an interrupt is received. Interrupt mode IM2 allows indirect call to an interrupt service subroutine at a location specified by two bytes: the contents of the I-register and the 8-bit word received from the interrupting device. This feature allows vectored interrupts servicing several peripheral devices whose subroutines may be different.

Interrupts. An interrupt is a process in which your computer stops executing the main program, and begins executing another program located somewhere else in memory. This is not a mere "JUMP" or "CALL" operation, but a response to an external stimulus.

There are several reasons why an interrupt capability may be required. One of these, for example, is the case of an alarm condition. We could, for example, use a computer in an environmental control system, and use the external interrupt capability to allow response to alarm situations (e.g., smoke detector, liquid level, burglar alarm, oven temperature, etc.). The computer would ordinarily go about some other chore, perhaps the business of controlling the system. But once during the execution of each instruction of the program, the CPU will interrogate the interrupt system. It is, then, monitoring the alarm status while executing some unrelated program. When an interrupt is received, indicating an alarm status, the computer would jump immediately to the program that services the interrupt: rings a bell, calls the fire department, turns on a light or sighs heavily.

Another application is to input data that occurs only occasionally, or whose periodicity is so long as to force the computer to do nothing for an inordinate amount of time. A real time clock, for example, might want to update its input to the computer only once per second or once per minute. An analog-to-digital converter (ADC) might have a 20 millisecond conversion time. Even the slower version of the Z80 CPU (using a 2.5 MHz clock) can perform hundreds of thousands of operations while waiting for the ADC to complete its conversion job. Since the ADC will not provide valid data until after the expiration of the conversion time, waiting for those data would be a tremendous waste of CPU time.

Another use is to input or output data to/from a peripheral device such as a line printer, teletypewriter, keyboard, CRT terminal and so forth. Those electromechanical devices are notoriously slow to operate. Even so-called "high speed" line printers are considerably slower than even the slowest Z80 CPU. A classic example is the "standard" 100 word per minute teletypewriter. A "word" in this case is five ASCII characters, so we have to output 500 characters per minute to operate at top speed. This rate is eight characters per second, so each character requires 1/8 second, or 125 milliseconds, to print. The CPU, on the other hand, is a trifle faster. It can output the character to the input buffer of the teletypewriter in something like three microseconds. The Z80 can execute almost 42,000 outputs in the time it takes the teletypewriter to print just one character.

There are at least two ways to handle this situation, and both involve having the peripheral device signal the CPU when it is ready to accept another character. This job is done by using a strobe pulse from the peripheral, issued when it is ready to receive or deliver another data byte. One way to handle this problem is have the programmer write in a periodic poll of the peripheral. The strobe pulse is applied to one bit of an input port. A program is written that periodically examines that bit to see if it is HIGH. If it is found to be HIGH, then the program control will jump to a subroutine that services the peripheral. But this approach is still wasteful of CPU time, and places undue constraints on the programmer's freedom.

A superior method is to use the computer's interrupt capability. The peripheral strobe pulse becomes an interrupt request. When the CPU recognizes the interrupt request, it transfers program control to an interrupt service subroutine (i.e., a program that performs some function required for the operation of the peripheral that generates the interrupt). When the service program is completed, then the control is transferred back to the main program at the point where it left off. Note that the CPU does not recognize an interrupt request until after it is finished executing the current instruction. Program control then returns to the next instruction in the main program that would have been executed had the interrupt not occurred.

Types of Z80 Interrupts. There are two basic types of interrupt recognized by the Z80 CPU: nonmaskable and maskable. The nonmaskable interrupt is executed next in sequence regardless of any other considerations. The maskable interrupts, however, depend upon the condition of an interrupt flip-flop inside of the Z80. If the programmer wishes to mask, i.e., ignore, an interrupt, then the appropriate flip-flop is turned off.

There are three distinct forms of maskable interrupt in the Z80, and these take the designations Mode-0, Mode-1 and Mode-2. There are two interrupt input terminals on the Z80 chip. The NMI (pin 17) is for the nonmaskable interrupt, while the INT is for maskable interrupts.

The nonmaskable interrupt (NMI) is much like a restart instruction, except that it automatically causes program control to jump to memory location 0066H, instead of to one of the eight standard restart addresses. Location 0066H must be reserved by the programmer for some instruction in the interrupt service program, very often an unconditional jump to some other location higher in memory.

The Mode-0 maskable interrupt causes the Z80 to pretend that it is an 8080A microprocessor, preserving some of the software compatibility between the two CPU's. During a Mode-0 interrupt, the interrupting device places any valid instruction on the CPU data bus, and the CPU executes this instruction. The time of execution will be the normal time period for that type of instruction, plus two clock pulses. In most cases, the interrupting device will place a restart instruction on the data bus, because all of these are one-byte instructions. The restart instructions transfer program control to one of eight page-0 locations.

Any time that a RESET pulse is applied (i.e., pin 26 of the Z80 is brought LOW), the CPU automatically goes to the Mode-0 condition. This interrupt mode, like the other two maskable interrupt modes, can be set from software by executing the appropriate instruction (in this case an IM0 instruction).

The Mode-1 interrupt is selected by execution of an IN1 instruction. Mode-1 is totally under software control, and cannot be accessed by using a hardware action. Once set, the Mode-1 interrupt is actuated by bringing the INT line LOW momentarily. In Mode-1, the Z80 will execute a restart to location 0038H.

The Mode-2 interrupt is, perhaps, the most powerful of the Z80 interrupts. It allows an indirect call to any location in memory. The 8080A device and the Z80 operating in Mode-0 permits only eight interrupt lines. But in Mode-2, the Z80 can respond to as many as 128 different lines. Mode-2 interrupts are said to be "vectored," because they can be made to jump to any location in the 65,536 bytes of memory.

THE 6502 MICROPROCESSOR

Our next example of an eight-bit microprocessor is the 6502 device, and its related family of microprocessors. The 6502 forms the heart of the Apple *II*e, Franklin ACE, KIM-1, SYM-1, and AIM-65 microcomputers. Related chips are used in other microcomputers. For example, the 65C02 (a CMOS version of the 6502) is used in the Apple *II*c, the 6510 is in the Commodore 64, etc. Because of its use in these popular machines, it is at least as important as the Z80.

The 6502 is also worth considering because it uses a different design philosophy than the Z80. Where the Z80 is register oriented, the 6502 is memory oriented. Where the Z80 contains a large array of internal registers, the 6502 has but a few "standard" registers. Where the Z80 instruction set contains unique I/O instructions, the 6502 is memory-mapped (i.e., I/O ports are addressed as if they are memory locations); hence, an input operation on 6502 is only a memory read, while an output operation is a memory write. Although it seems that the advantages are on the side of the Z80, a large number of designers selected the 6502. There are cases, incidentally, where a 2 MHz 6502 will execute a benchmark program faster than a 4 MHz Z80 machine.

The 6502 microprocessor chip is available from more than fifteen different sources. Figure 17-5 shows the block diagram of the 6502 architecture, while Fig. 17-6 shows the pinouts.

That there are certain similarities between the Z80 and 6502 devices testifies only to the fact that they are both microprocessor chips. The Z80 and 6502 devices are designed with different philosophies, which are reflected in their respective internal architectures. The Z80 allows separate input/output commands. The lower order byte of the address bus will carry the port number (256 ports, numbered 000 - 255) address, while the high order byte of the address bus carries the contents of the accumulator that are to be output. The I/O data is fed to and from the accumulator over the data bus. The 6502, on the other hand, uses a memory mapping technique in which each I/O port is designated as a separate location in memory. We can then read or write to that memory location, depending upon whether the operation is an input or output. The 6502 also lacks the multiple internal registers of the Z80. But this feature, like the lack of discrete I/O ports, does not hinder most microcomputer designs. Very few microcomputers will need more than a total of a dozen or so I/O ports and/or registers. Also, very few microcomputers will need the entire 64K of available memory addresses. In fact, many systems have less than 48K of memory. This allotment of memory would leave all locations above 48K for "firmware" (i.e., ROM) programs and I/O port or register selection.

Blocks in Figure 17-5 that have similar names to blocks in the Z80 diagram perform roughly similar (sometimes identical) jobs for the 6502 device. Note, however, that the Program Counter (PC) is divided into two eight-bit (one byte) registers called PCL (for low order byte) and PCH (for high order byte of the 16-bit address). Similarly, the address bus is divided into low (ADL) and high (ADH) order segments. Unlike the Z80, the 6502 uses a multi-phase clock for timing of the operations. The 6502 pinouts shown in Figure 17-6 are defined below. Some of them are similar to Z80 pinouts, while others are unique to the 6502.

DBO - DB7	Eight bit data bus
ABO - AB15	Sixteen bit address bus
P_0	CPU clock
P_1	Phase-1 system clock
P_2	Phase-2 system clock
R/W	Read/Write (HIGH for read, LOW for write)
IRQ	Active-LOW input is used to interrupt the program being executed so that a subroutine can be executed instead. This interrupt line is active only if the internal 6502 interrupt flip-flop is enabled.
NMI	Nonmaskable interrupt is similar to IRQ, except it cannot be ignored.
RESET	Active-LOW reset line. This line is essentially a hardware jump to a location in page-F, where the location of the first instruction in the program. Can be used for either manual or power-on reset operations.
RDY	The active-HIGH ready signal inserts a wait state

Fig. 17-5. 6502 architecture (courtesy of Mostek).

into the normal cycle process. The RDY line must make a negative-going (H-to-L) transition during the P_1 HIGH clock cycle during any operation other than a write.

SO   Active-HIGH set overflow input. It will allow hardware setting of the internal overflow flag if it makes a negative-going transition during the trailing edge of the P_1 clock cycle.

SYNC   Active-HIGH output that is used to indicate the instruction fetch machine cycle.

ABO-AB15

A8-A15 — ABH — SYSTEM ADDRESS BUS

A0-A7 — ABL

RDY RES IRQ MNI

R/W SO SYNC — INTERRUPT CONTROL LOGIC ← TIMING CLOCK

INSTRUCTION DECODER AND CONTROL LOGIC

$\emptyset_2$ $\emptyset_1$ $\emptyset_0$

INSTRUCTION REGISTER

PROCESSOR STATUS REGISTER

DATA BUS

6502 Processor Status Register

Like the F and F' registers in the Z80, the 6502 has a status register that can be used by the programmer in a number of important ways. The status register can be read by the program during certain operations, but is inaccessible to external hardware. The register contains eight bits defined as follows:

Bit	Use
0	Carry or "C-flag." Indicates that a carry occurred from the

Fig. 17-6. 6502 pinouts.

Bit	Use
	accumulator on the last instruction executed. Active-HIGH.
1	Zero or "Z-flag." Indicates that the last instruction executed resulted in a zero (Z = 1), or nonzero (Z = 0) answer.
2	(unused)
3	Decimal mode (D-flag). When this bit is HIGH it causes a decimal add with carry or subtract with borrow for BCD arithmetic operations.
4	Break (B-flag). HIGH when an interrupt is executed.
5	(unused)
6	Overflow or "O-flag." Active when overflow occurs.
7	Sign or "S-flag." Indicates whether an arithmetic result is negative or positive.

6502 Timing and Control Signals

The 6502 has relatively fewer control signals compared with

the Z80 chip. There are complementary Phase-1 and Phase-2 clock signals (Fig. 17-7), which are derived from 1 MHz (2 MHz in some versions) clock signal either generated by the 6502 or presented at the clock input. The main difference in this respect between the Z80 and 6502 machines is that the 6502 has a single read/write line designated "R/W." This signal is HIGH for read operations and LOW for write operations. In both cases, it is during Phase-2 of the master clock cycle that these signals become valid.

A read operation timing diagram is shown in Fig. 17-8. The Phase-1 period is a time of set-up. During this period a valid address is stabilized on the address bus (A0 - A15), and the R/W control signal goes HIGH. This operation commences a couple hundred nanoseconds into Phase-1 period, and is completed by the onset of Phase-2. Thus, when the computer enters Phase-2, all is ready. When the Phase-2 clock goes HIGH, the commanded action will take place. A three-input NAND gate can be used to create a system READ signal (or IN signal, if the memory location is an input port): one input looks at Phase-2, one input looks at R/W and one input at an active-HIGH device SELECT signal. These signals are covered in more detail in Chapter 18. The confluence of these three active-HIGH signals determines the input and/or operation.

The timing diagram for a write operation is exactly like the read operation shown in Fig. 17-8, except that the R/W line goes LOW during Phase-1 and remains LOW throughout Phase-2. During the last 300 nanoseconds of Phase-2 the data is placed on the data bus (DB0 - DB7); the data remains valid until after Phase-2 goes LOW again.

In order to generate a system WRITE or OUT signal we can invert the R/W signal and then use the same type of three-input NAND gate as described above.

6502 Interrupt Modes

An interrupt line is an external control input that is used to grab control of the processor. When the interrupt lines goes active

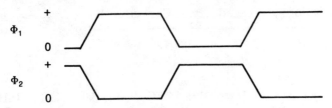

Fig. 17-7. 6502 clock cycle.

Fig. 17-8. 6502 read/write cycle.

(LOW), the control shifts to another portion of the program. There are three forms of interrupt line on the 6502: reset (RST), interrupt request (IRQ) and nonmaskable interrupt (NMI); these are analogous to interrupts of similar name on the Z80 device.

The RST line differs from its Z80 counterpart because it is vectored. On the Z80 machine the reset line causes a hardware JUMP to location 0000H. On the 6502, however, the JUMP occurs to a vectored location. The exact location of the reset is determined by the contents of memory locations FFFC(H) and FFFD(H). The low order byte of the destination address is stored in FFFC(H), while the high order byte is in FFFD(H). For example, consider the following:

Location	Hexadecimal Contents (H)
FFFC(H)	08(H)
FFFD(H)	A0(H)

We have 08(H) stored in FFFC(H), and A0(H) in FFFD(H). When the RST line is brought LOW, the control of the program jumps to the program subroutine that begins at location A008(H) in memory. This is an example of a vectored reset.

The interrupt request (IRQ) line is also vectored, with the jump-to location stored at FFFE(H) and FFFF(H). There are two conditions on the interrupt request. The IRQ is maskable because it will respond only if the I-flag in the 6502 processor status register is zero. This condition can only be produced by a deliberate programming step—the CL1 instruction clears the I-flag. The second condition is that the BRK instruction uses the same vector locations as the IRQ. If a "break" occurs, the B-flag of the PSR will be HIGH. Thus, the subroutine pointed to by FFFE(H) and FFFF(H) must contain a short test to check the status of the B-flag.

The nonmaskable interrupt (NMI) does not depend upon the condition of any internal condition, and cannot be ignored by the 6502. The vector addresses for the NMI instruction are located at FFFA(H) and FFFB(H).

Chapter 18

Memory-I/O

Interfacing in Computers

In most digital computers or other instruments, busses are used to transfer data or provide addresses to locations. These busses are parallel data paths between sections of the computer. But because all sections share the same bus, some means must be provided for decoding specific input/output port or memory location addresses. There are, in most eight-bit microcomputers, 256 different allowable I/O ports, and 65,536 memory locations! Obviously, in order to avoid chaos, we must tell the memory:

- ☐ That a memory operation is taking place.
- ☐ Whether it is a *read* or *write* operation.
- ☐ Which of 65,536 locations is intended in this operation. (The first two steps are combined in some microprocessors.)

Similarly, for I/O operations, we need to do the same thing, except that the first step is modified to indicate an I/O operation:

- ☐ Specify one of 256 possible (000 to 255) devices or I/O ports is intended.
- ☐ Indicate to the I/O ports than an I/O operation is to take place.
- ☐ Specify whether the operation is an input (read) or output (write).

A *device selector* is used to turn on, or signal, some device or circuit external to the CPU. Basically, most (perhaps all) such devices are seen by the CPU as either *memory* or *I/O ports*. The device select function, then, is essentially the same as these functions.

BINARY WORD DECODERS

Binary word decoder circuits are designed to provide an output indication when a specified binary word is applied to its input. These circuits are used to decode the data on the address bus to determine which of 65,536 memory locations is intended, or which of 256 I/O ports are intended.

In simple, four-bit applications we can use some of the decoder and data distributor ICs in the 7400-series (TTL) line. We may also use the 7485 four-bit comparator.

Figure 18-1 shows the use of 7442 1-of-10 BCD decoder. This device is normally used to select the vertical column decade light, or *Nixie*® tube elements that correspond to the four-bit (BCD) binary word at the inputs. Note that the normal range of four-bit binary is 0 to 15 decimal. This means that only part of the range is usable; i.e., 0 to 9. Within that constraint, however, we can find

Fig. 18-1. Four-bit decoder using 7442 device.

much application for the 7442 device. When the selected binary word is applied to the inputs (B0- B3), then the corresponding output drops LOW. As an example, let us say that we want to designate a given device or circuit as input device No. 6 (6_{10} = 0110_2). Since 0110 is within the range of the 7442, we can use that IC instead of some more expensive device. In our simplified example, we will use an LED (light-emitting diode) to simulate a peripheral device that wants to see a LOW on the *enable* (EN) input to turn on. The LED remains turned off if any binary word other than 0110 is applied to B0-B3. But if 0110 is seen, then output No. 6 (pin No. 7) of the 7442 goes LOW, turning on the LED.

The use of the 7485 data comparator is shown in Fig. 18-2. This TTL device examines two, four-bit binary words (designated A and B), and issues outputs indicating one of three possible states: A equals B, A less than B, A greater than B. In this application, we are interested in A = B. Note that there are also three inputs for the same three conditions. We tie A > B and A < B to ground and A = B to +5VDC (logical-HIGH). We will program the B inputs with the number in binary (0000 to 1111, or 0 to 15) assigned to the device. If, as in the previous example, we call the device port No.

Fig. 18-2. Programmable decoder using the 7485 device.

Fig. 18-3. Four-bit 1-of-16 decoder using 74154 device.

6, we will load 0110 into 7485 input B. This job is accomplished in the example circuit of Fig. 18-2 by switches, but would probably be hand-wired in most equipment. When 0110 is applied to the A-inputs, the A = B condition exists, so pin No. 6 goes HIGH. In this circuit, the select pulse output is active-HIGH, so will turn on only those devices that want to see a positive-going enable pulse. If an active-LOW enable pulse is needed, then an inverter at the 7485 A = B output will be required.

A one-of-16 decoder is shown in Fig. 18-3. This circuit is based on the 74154 TTL data distributor IC. By changing the level applied to pin No. 18, we can determine whether the selected output is active-HIGH or active-LOW. A HIGH on pin No. 18 produces the active-HIGH mode, while a LOW on pin No. 18 produces the active-

311

LOW condition. The input is four-bit binary, allowing up to 16 counts. Corresponding to the 16 unique states in four-bit binary (0000 to 1111). The device enable line on the peripheral will be connected to the 74154 output corresponding to the 16 unique states in four-bit binary words (0000-1111). The 74154 also allows selection of active-HIGH/active-LOW output; not possible on the 7442. The 74154 is a 24-pin DIP device and requires somewhat more space on a printed circuit board than does the 7442.

Thus far, our discussion has centered around four-bit data words. But this is legitimate only when less than sixteen choices are needed; i.e., when all addresses can be given in the number range of the four-bit word (less than 16 choices). But most microcomputers, and those digital instruments that are based on the microprocessor chip, are oriented around eight or 16 bit data and address buses. If we wish to take full advantage of these devices, then we must be able to decode word lengths longer than four bits.

EIGHT-BIT WORD DECODERS

An eight-bit binary word decoder circuit used in many microcomputer designs is shown in Fig. 18-4. This circuit is based on the 7430 eight-input TTL NAND gate IC. Recall the operation of a NAND gate: if any one input is LOW, then the output is HIGH. It requires all inputs (in the case of the 7430, all eight) must be HIGH the output to be LOW. In this type of decoder, we want to connive to make all eight inputs of the NAND gate HIGH when the required word appears on bits A0 through A7 of the address bus. But there is only one word (1111) that has all HIGH bits natu-

Fig. 18-4. Eight-bit decoder.

Fig. 18-5. Decoder for the word 11010011.

rally. We must use inverters to flip the normally LOW bits of the desired word to HIGH. Only two inverters are shown here (at the LSB and MSB positions), but you may assume as many as eight, if needed.

Let's consider the A0 inverter. If the bit at that position in the desired binary word will be HIGH, then switch S0 will be in the "1" position. The A0 line is now connected directly to one input of the NAND gate. But what happens when the anticipated A0 bit is LOW? Here we must set switch S0 to the "0" position. Line A0 is not connected directly to an input of the NAND gate, but is first inverted; A0 is applied to the A0 input of the NAND gate. If all switches are correctly set, then the nand gate will see the binary word 11111111 *only* when the correct word is applied to A0-A7.

Figure 18-5 shows an actual example of a decoder. This particular circuit could be used to decode I/O port addresses (in many microcomputers the eight-bit I/O port address appears on the lower byte of the address bus, hence the A0 through A7 designations). This circuit is set to decode the binary word 11010011. Bits A0, A1, A4, A6, and A7 will be "1" (HIGH), so those lines are connected directly to the 7430 NAND gate. Bits A2, A3 and A5 will be "0" (LOW) so they must be inverted prior to being applied to the 7430 inputs. The only binary word an A0- A7 that will produce 11111111 at the 7430 inputs, then, is 11010011, the required word.

The circuits of Figs. 18-4 and 18-5 have proven extremely popular with designers of digital equipment. They can boast both low cost and simplicity.

An eight-bit circuit based on the 7485 TTL comparator IC is shown in Fig. 18-6. The 7485 is a four-bit device, so two must be

313

Fig. 18-6. Cascading 7485 devices to make eight-bit decoder.

used in cascade to accommodate eight-bit situations. This is done by connecting the outputs of one stage to the *cascade inputs* of the following chip. IC1 will not output a HIGH unless the A = B criterion is met by both chips.

The operation of this circuit is essentially the same as in the four-bit version. The desired address is loaded into one set of inputs, while the bus being monitored is connected to the other. Pin No. 6 goes HIGH *only* if A = B. At all other times pin No. 6 is LOW.

An extension of the 7442 circuit is shown in Fig. 18-7. Recall that the 7442 is a BCD-to-1-of-10 decoder. It will produce a LOW output that corresponds to any decimal equivalent of binary from 0000 to 1001 (0 to 9 in decimal). By using two 7442 devices and a single, two-input, 7402 NOR gate we can generate many more combinations (00 to 99). The inputs to the 7442 are connected in four-bit "nybbles" (half a byte) to the eight-bit bus. The inputs to the 7402 are connected to the output corresponding to the desired code. Assume that IC1 is the low order half-byte, and that IC2 is

Fig. 18-7. Selectable eight-bit decoder.

Fig. 18-8. Operation of the 7442 circuit of Fig. 17-7.

the high order half-byte. Let us further assume that we assign output port No. 26 to this circuit (see Fig. 18-8).

When the code on the address bus is 00100110, then the "2" output of IC2 and the "6" output of IC1 will both drop LOW simultaneously. Recall the rules governing a TTL NOR gate. The output is HIGH only if both inputs are LOW. This condition, then, generates a select pulse at the output of the NOR gate.

I/O SELECT CIRCUITS

Selecting an input/output port requires more than an address decoder. The address bus is also used for memory addressing, so to avoid confusion, we must be able to recognize the I/O state of the CPU. In this section, we will describe typical circuits using the Z80 microprocessor IC control signals. Keep in mind that other microprocessor chips will use slightly different control signals, or will use different names for the same signals.

The Z80 uses two separate pins for read and write signals. The active-LOW outputs are designated $\overline{RD}$ and $\overline{WR}$, respectively. But the $\overline{RD}$ and $\overline{WR}$ terminals are also used for memory operations, so we also need another set of control signals to differentiate between these functions. In the Z80 device, there are separate *input/output request* (IORQ) and *memory request* (MREQ) signals to service that function. When an I/O operation is to take place, the IORQ goes LOW. But if a memory operation is taking place instead, then MREQ goes LOW.

Figure 18-9 shows a circuit that will generate OUT and IN directional commands when the desired I/O port address is present on the lower eight bits of the address bus. The circuit is built around the 7402 device, a TTL NOR gate. The basic operating rules of the NOR gate are: both inputs of the NOR gate must be LOW in order to generate a HIGH output. The output will be LOW if either NOR gate input is HIGH. Gates G1 and G2 can only be ena-

Fig. 18-9. Typical WR/RD circuit to generate IN and OUT commands.

bled when IORQ is LOW (active) and the appropriate output (inverted) is received from G3 and G4, respectively.

Let us assume an *output* operation is to take place, and the correct port address has been received at the address decoder. The correct address will cause the SELECT line to drop LOW, while an output command causes WR (an output is a write operation) to drop LOW. Both inputs of G3 are now LOW, so its output goes HIGH. This condition forces the output of inverter NO. 1 LOW. Gate G1 now sees both inputs LOW because IORQ is also LOW at this time. The output of G1, then, snaps HIGH, generating the OUT signal.

Similar reasoning describes the creation of the IN signal; only now the operative gates are G2 and G4. Gate G2 is enabled when the IORQ lines goes LOW, indicating that an I/O operation is taking place. But the IN signal cannot be created until the RD line goes LOW (inputs are read operations). When both criteria are satisfied, then both inputs G4 are LOW, and that makes the output of G4 HIGH. Now all three criteria for an input operation are satisfied, we can generate an IN signal. The output of gate G2 goes HIGH.

We may also use the 7442 device to generate an IN or OUT signal. A sample circuit is shown in Fig. 18-10. The 7442 BCD-to-1-of-10 decoder causes one of ten possible outputs to drop LOW in response to an appropriate four-bit binary word at the input. We can make I/O control signals of the Z80 look like a three-bit word. The fourth bit required by the 7442 is obtained by grounding the D input (pin No. 12) of the 7442. This makes the most significant bit of the code always LOW (logical-0). To generate a binary code, we connect the Z80 to the 7442 as follows:

7442 Input	Z80 Control Signal
A	IORQ
B	RD
C	WR
D	(Zero)

The respective binary codes, then, will be:

Command	BCD	Decimal	7442 Pin No.
IN	0100	4	5
OUT	0010	2	3

We must also account for the I/O port address, as indicated by either SELECT or $\overline{\text{SELECT}}$ from the address decoder circuit. There are at lest three common approaches to solving this problem. One of them is to use NOR gates, as shown in Fig. 18-10. The other two involve using the D-input of the 7442.

In Fig. 18-10, two NOR gates are used. One input will go LOW when $\overline{\text{SELECT}}$ from the address decoder goes LOW. The remaining pinouts are connected to the 7442 outputs that correspond to IN and OUT. The operation is as follows:

Input operations:
☐ $\overline{\text{IORQ}}$ and $\overline{\text{RD}}$ go LOW, so the BCD word at the 7442 input

Fig. 18-10. Interfacing Z80 with 7442 decoder.

is 0100. This causes output No. 4 (pin 5) to go LOW. This pin 5 signal is the $\overline{IN}$ signal.

☐ At the same time, the port address appears on bits A0 through A7 of the address bus, and the address decoder issues a $\overline{SELECT}$ signal.

☐ With both $\overline{SELECT}$ and $\overline{IN}$ present at its inputs, the output of G1 goes HIGH. This becomes our IN signal.

Input operations:

☐ $\overline{IORQ}$ and $\overline{WR}$ go LOW, and the port address appears on bits A0 through A7 of the address bus.

☐ The input of the 7442 sees 0010, so the No. 2 output (pin 3) goes LOW. This is the $\overline{OUT}$ signal.

☐ At this time, $\overline{SELECT}$ goes LOW.

☐ Both inputs of gate G2 are now LOW, so its output goes HIGH to produce the OUT signal.

Figure 18-11 shows a method based on the 7442, but without the NOR gates. Here we are using the D-input to detect the $\overline{SE-LECT}$ output of the address decoder. We still use the codes 0100 and 0010 for IN and OUT, respectively, but they are now generated without the NOR gates.

If a SELECT signal is available, then the output of the address decoder goes HIGH for the correct address. This would generate the codes 1100 and 1010 for $\overline{IN}$ and $\overline{OUT}$, both illegal on the 7442. To overcome this slight hitch, we must pass the select signal

Fig. 18-11. Making better use of the D input of the 7442.

Fig. 18-12. Inverted SELECT to make SELECT signal.

through an inverter to create a $\overline{\text{SELECT}}$ signal. This is done in Fig. 18-11. See also Fig. 18-12.

If we wanted to use 74154 binary-to-1-of-6 decoder, however, we could do substantially the same thing, but without the inverter. One might wonder, however, if that is economical.

One last approach to creating I/O capability is to use one of the complex I/O function ICs offered by microprocessor manufacturers. These chips are usually called peripheral interface chips, or something similar (PIO, PIA, etc.). Zilog, Inc., the prime manufacturer of the Z80-family of devices, makes at least two types: serial (Z80-SIO) and parallel (Z80-PIO). Figure 18-13 shows the Z80-PIO parallel interface chip connected to the Z80 CPU chip. Note that certain pinouts on the PIO have the same designations as pinouts on the Z80 chip. This means that they are to be connected to the designated pins on the Z80 for proper operation. The Z80-CPU and Z80-PIO are, then, a matched pair designed to have like pins connected together. In this application we also pair $\overline{\text{IORQ}}$ and $\overline{\text{CE}}$ (chip enable), on the theory that the Z80-PIO need not be enabled unless an input/output operation is taking place, and the $\overline{\text{IORQ}}$ terminal indicates the I/O state. You may wonder why one would have a separate chip enable terminal at all. The reason is that the PIO may be used in a circuit where there is more than one type of I/O.

INTERFACING MEMORY

One of the first accessories one adds to a computer is external memory. Although some of the latter CPU chips contain on-board memory locations, most require external memory before they can be used to do any real work. At one time, "memory," consisted of tiny doughnut-shaped ("toroidal") cores of ferrite material. These

Fig. 18-13. Interfacing I/O using the Z80 PIO.

were magnetized to store data. The circuitry required to support core memory was tremendous by today's standards. In modern computers we find solid-state memory, which requires little in the way of support circuitry. Solid-state is vastly easier to support, is a lot cheaper, and much faster operating. Core memory, on the other hand, was nonvolatile; i.e., you don't lose the data just because the power supplies goes off. Many solid-state memories using the popular bipolar transistor technology, unfortunately, *are* volatile; they seem to suffer an, errr, uhhh, errr, a whatacallit, *amnesia* on power down. Recent solid-state memory devices, however, have partially solved that problem by using MOS technology. A small bias voltage, at almost zero current, will sustain memory even though power is off. This bias can be obtained from a small battery. You may have already seen this technology in some consumer devices. Some calculators and electronic checkbooks (a form of calculator) boast continuous memory. These machines remember the last data (your present checking account balance) present in the registers at power turn/off. At least one television receiver uses electronic memory for their programmable TV tuner. The customer would program 10 to 15 registers with the TV channels present in the locale, and thereby custom tunes the set. A small battery is used to keep the data current, even though the set is turned off, or unplugged. CMOS, PMOS and NMOS memory requires so little current that the memory hold batteries are said to last two years!

Read only memory, or ROM, memory chips are pre-programmed (permanently or semipermanently in the case of

320

EPROMs). The CPU cannot write data into a ROM, but can read data *from* memory; hence, the name *read only memory* (ROM).

Figure 18-14 shows how to interface a certain class of ROM device. This case covers those devices with two chip enable inputs. CE1 is active-HIGH and CE2 is active-LOW. Input CE2 is connected to A10 of the address bus. Input CE1, however, requires two conditions that must be met before it can become active: $\overline{RD}$ and $\overline{MREQ}$ must both be LOW, indicating that a memory read operation is being executed by the CPU.

In the simplified case shown, we use bit A10 to turn memory on and off. Here we are designating the 1024 addresses of the ROM as the lowest 1024 bytes of the allowable 65,536 byte memory space. A0 through A9, then, are active and A10 can be used as a chip enable, freeing the rest of the available addresses for random access read/write memory (RAM). Bit A10 of the address bus goes HIGH, disabling the ROM, for those addresses higher than 1024.

An alternate system, for use with ROMs and EPROM that store only 256 bytes (like the 1702) is shown in Fig. 18-15. This device uses only eight address bits (A0 through 7) and has a single active-LOW chip enable input.

We want the $\overline{CE}$ to go LOW only when three conditions are met:

☐ $\overline{MREQ}$ (memory request) is LOW.
☐ $\overline{RD}$ (read) is low.
☐ A8 is LOW.

The first of these three ($\overline{MREQ}$ and $\overline{RD}$) are the Z80 control signals that together indicate a read from memory operation is taking place. The third tells us that the address on the address bus is in the lower 256 bytes of memory space. All other possible addresses has bit A8 HIGH.

Fig. 18-14. Using MREQ and RD signals to interface ROM.

We can accomplish this job using a three-input NOR gate. Its output stays LOW as long as any *one* input is HIGH. It will go HIGH only when all three inputs are LOW. This happens in Fig. 18-15 only during memory read operations in the lower 256 bytes of memory. The NOR gate output must be inverted (it is active-HIGH) before it will properly drive the active-LOW input of the EPROM ($\overline{CE}$).

You may be confused by our seeming use of NOR gates for apparently NAND functions. In the case above, the output is HIGH only for the condition $\overline{MREQ}$ AND RD AND $\overline{A8}$. This is merely terminology: a positive logic NOR gate is a negative logic NAND gate (and vice versa). TTL and most CMOS gates are assigned names that assume positive logic. In this circuit, however, we have active-LOW chip enable functions, so the logic is negative. This is indicated by the "NOT bar" symbols over MREQ, RD, and A8.

Figure 18-16 shows an alternate form of the circuit in Fig. 18-15. Here we are not using a triple input NOR gate IC, but instead are making a three-input NOR function by properly interconnecting an inverter to a pair of two-input NOR gates. This does not, at first blush, appear to be good economics; it requires three gates extra. But under the right circumstances, digital equipment makers will opt for this type of circuit because it *is* economical. Keep in mind that typical ICs in both TTL and CMOS are *hex* (6) inverters and quad (4) two-input gates. If there are inverter and NOR gates used elsewhere in the circuit, then there might well be a sufficient number of spare inverter/NOR-gate sections to make the circuit of Fig. 18-16.

Alternatively, what if there is no spare inverter available? We would need an IC with a three-input NOR and another with six inverters. This results in five wasted inverted sections. But a single quad, two-input, NOR gate will suffice for Fig. 18-16 if two of the

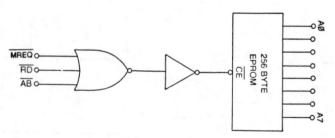

Fig. 18-15. Interfacing the 256-byte ROM.

Fig. 18-16. Using discrete gates.

NOR gate sections were inverter-wired (see two examples in Fig. 18-17).

Thus far, we have considered only the simplest cases; i.e., ROM/EPROM in the lower portion of memory. But as man does not live by bread alone, computers do not live on 1k bytes of ROM alone. In the next section we will consider random access read/write memory to 64k bytes (65,536).

RAM R/W MEMORY ORGANIZATION

Most computers use arrays or banks of memory ICs, each containing a fraction of the total memory. These are connected with appropriate decoding to the address bus and directly to the data bus.

Consider Fig. 18-18. Here we see the popular 2102 memory ICs, which are 1024X1 bit devices. This means that the address bus can access 1024 1-bit locations. To make a 1024 bank of one byte locations, then, requires eight ICs. A total of 64 devices are required to make a 64K array of memory locations. This circuit is actually somewhat simplified in that only one data bus is shown. The actual 2102 devices has *data to CPU* and *data from CPU* terminals, so a little external logic is required to merge these into one data bus.

Another organization scheme is shown in Fig. 18-19. In this case, two ICs are paired, each able to store 256 four-bit words. Two

Fig. 18-17. NOR gate inverter connected.

Fig. 18-18. Interfacing to RAM organized in 1024X1 bit chips.

devices together, then, will store 256 bytes. Also shown in this figure is the RD and MREQ/A1Ø logic required to turn on this bank of memory.

MEMORY BANK SELECTION

When memory is organized into blocks or banks, it becomes

Fig. 18-19. Interfacing RAM organized in 256X4 chips.

324

Fig. 18-20. Generating block select function.

necessary to be able to select the required bank and the specific address within that bank. Figure 18-20 shows a bank select scheme for eight-bit microcomputers capable of addressing 8K bytes of memory in 1K byte blocks. Once again, we see the use of the 7442 BCD-to-1-of-10 decoder IC.

Bits A10 through A9 of the address bus are sufficient to address 1024 locations in memory. The pins for these bits are all tied in parallel such that all A0 pins are tied together, all A1 pins are tied together and so on, forming an address bus. But unfortunately, there is not much use in addressing the same location in all eight blocks simultaneously. We need a method for determining which block of 1024 bytes is intended, and then to turn on the CE terminals of those chips. We start by examining bits A10 through A13 (see Fig. 18-21). Note that bit A13 is always LOW, so it can be ignored. If we assign the eight blocks the number 0 through 7, we find that bits A0 through A12 form a binary number equivalent to the block number. We may now use A10 through A12 as the "1-2-4" inputs to a 7442. The "8" input of the 7442 (pin No. 12) can be grounded because A13 is always zero.

We now have a circuit that is capable of selecting the correct location in the correct block for the applied address. The 7442 outputs are active-LOW, so will interface directly with most common memory IC devices.

For computers that allow addressing of more than 8K bytes of memory, we must extend this concept a little. Figure 18-21 shows

Fig. 18-21. Generating bank and block select.

a circuit using eight 8K byte banks (each like Fig. 18-20) to provide a complete 64K memory. Each eight kilobyte bank has its own 7442 to sense the sate of bits A10 through A12 . This scheme is identical to Fig. 18-14A. Similarly, bits A0 through A9 of all memory devices are bussed together.

The 64K circuit of Fig. 18-21 uses one additional 7442 decoder to examine bits A13 through A15. The "8" input of the bank select 7442 is grounded permanently, but the "8" inputs of all eight-block select 7442 are used as chip enable pins. Remember, since only eight possibilities (0 through 7) are allowed, the "8" input is LOW for all valid addresses. The block select 7442s, then have their

Table 18-1. Condition Codes at Input of 7442.

		A13	A12	A11	A10
Block-0	0-1K	0	0	0	0
Block 1	1K-2K	0	0	0	1
Block 2	2K-3K	0	0	1	0
Block 3	3K-4K	0	0	1	1
Block 4	4K-5K	0	1	0	0
Block 5	5K-6K	0	1	0	1
Block 6	6K-7K	0	1	1	0
Block 7	7K-8K	0	1	1	1

"8" inputs connected to the appropriate outputs (0 through 7) of the bank select 7442.

Let us describe the conditions in the circuit of Fig. 18-21 if the desired memory address is 11613_{10}. This address, written in binary form, is 0010110101011101_2. This address can be divided up into location (bits A0 through A9), block (A10 through A12) and bank (A13 through A15):

We would expect this binary word to address location 349 (out of the possible 1024), in block 3 (out of 8) of bank 1 (of 8). This is at least "in the ball park" because bank 1 is 8-16K, and 11613_{10} falls within that range. We can check this further, however, by doing a little arithmetic:

```
      0101011101        349
    011-----------      3072
  001--------------     8192
  ----------------------------
  0010110101011101     11613
```

When 0010110101011101 appears on the address bus, location 349 in all 64 memory ICs are addressed simultaneously. The proper combination of block and bank select logic will turn on only the single set of eight ICs that contains location 11613. In this case, block No. 3 is selected, so all eight-block select 7442s see 011 on bits A10 through A12. But only the block select chip serving bank No. 1 will see the *correct* code 0011; all others see 1011. This is because output No. 1 of the bank select 7442 is LOW, while the others are HIGH; hence, we turn on bank No. 1, block No. 3, and location 349 to select location 11613.

Note that this is only one scheme of several that are commonly used in microcomputers. But this book must, because of space limitations, be representative rather than comprehensive.

MEMORY-MAPPED I/O

Using the regular input-output functions of any computer is

sometimes more limited, and time consuming, than the programmer might wish. All I/O instructions in most computers, for example, must be processed through the accumulator. In order to transfer data from a register (usually, but not always, the accumulator) to I/O, one must first pass through the accumulator, and then to the I/O port. Memory instructions, on the other hand, are often arranged to allow direct transfer between the accumulator and a location, another register and memory location, or from one memory location to another. These avoid the step of first loading the data into the accumulator before writing it to an I/O port. Also, another limiting factor is that there are only 256 different possible I/O locations, and 65,536 memory locations.

Memory-mapped I/O is a technique that allows us to treat peripherals and I/O ports as memory locations. This assumes, of course, that unassigned memory exists. Most microcomputers use less than 40K, and many less than 32K, of the available 64K. In those machines, it is a simple matter to add memory mapped I/O. If there is a full complement of memory (64K), then we can still use memory mapping only we must sacrifice the location to the peripheral. This schemes requires us to remember to not program or write data into, that location. Any data written to this location and to the peripheral assigned to that location.

Figure 18-22 shows a typical memory-mapped peripheral device. In this case, it is a digital-to-analog converter, or DAC (see Chapter 19). A DAC will produce an output voltage or current that

Fig. 18-22. Memory-mapped I/O (in this example using a DAC).

is proportional to the binary word applied to its input bits. We may, therefore, use a DAC to produce an output voltage for an oscilloscope, chart recorder, or analog control circuit under control of the computer. The binary word applied to the DAC is held in the eight-bit 74100 data latch. Data from the data bus is transferred to the latch outputs when line D is brought HIGH. The decoding scheme shown here is similar to other memory operations (again the Z80 uP chip is used as an example). Coexistence of $\overline{MREQ}$ and $\overline{WR}$ causes point A to go HIGH, which is inverted to force point B LOW. Point C, the remaining input of the NOR gate G2, goes LOW only when the memory address assigned to the DAC is present on the address bus. In any appropriate operation, where it is desired to send an output word to the DAC, both C and B are LOW, forcing D (the *strobe* terminal of the 74100 data latch) HIGH. Data on the D0 - D7 bus at that time instant is transferred to the latch output and held.

We may also use this same circuit, minus the DAC of course, as an output port. The outputs of the 74100 will then appear to the outside world as a TTL-compatible, eight-bit, binary output port.

An input port can be formed by using a tri-state bus driver in place of the 74100, and the $\overline{RD}$ signal instead of the $\overline{WR}$ signal. An extra inverter is also needed if the device uses an active-LOW chip enable (the usual case). Examples of typical circuits are shown in Figs. 18-23 and 18-24.

The circuit in Fig. 18-23 uses two quad buffers that have tri-state outputs (TTL device 74125). A "tri-state" device has three possible output states instead of the two normally associated with binary devices:

☐ Low impedance to ground (LOW).
☐ Low impedance to V + (HIGH).
☐ High impedance to both ground and V + (the output looks like a high-value resistor to the bus).

Tri-state devices can be connected to the data bus, and will not load the circuit unless the chip enable terminals (1, 4, 10, and 13 in the 74125 device) are LOW. If $\overline{CE}$ is LOW, then, the D0 through D7 outputs connected to the data bus will follow the B0 through B7 inputs. An inverter is required on the CE line to make it compatible with the circuit of Fig. 18-22. Of course, there might also be alternate logic schemes used that already provide an active-LOW chip enable function.

The above circuit uses the TTL 74125 device. Similar 2×4 bit circuits may also be used. One that comes immediately to mind is the Intel 8216 device. It is intended for the the 8080 microprocessor but works equally well in other applications.

An 8×1 bit circuit, capable of inputting an entire byte with one chip, is shown in Fig. 18-24. It uses the Intel 8212 device. This chip is also intended for service with the 8080 microprocessor but is often found in other applications. Like the circuit of Fig. 18-23, it is not a latch, but a tri-state buffer. When point D goes HIGH, the port is enabled. Data present at that instant on B0 through B7 is transferred to D0 through D7. When point D is LOW again, after the memory read command is executed, the outputs again go into the high-impedance state. Under that circumstance, the data bus doesn't even see the outputs of the 8212.

Fig. 18-23. Tri-state input port.

Fig. 18-24. Using Intel 8212 tri-state I/O device.

Both circuits of Figs. 18-23 and 18-24 can be used as an ordinary input port. We may also use the 74100 configuration of Fig. 18-22 as an output port. Modifications would be needed, but they are slight: replace $\overline{MREQ}$ with $\overline{IORQ}$ and use only A0 through A7 of the address bus.

Chapter 19

Data Conversion

We know that digital electronic circuits are happy only with *binary* signals. This means that the signals can take on only one of two possible values designated HIGH or LOW (or "1" and "0"). In between values will either be ignored, or will be accepted as one of the two permissible values. In TTL circuits, for example, HIGH is defined as + 5 volts and LOW is 0 V (grounded). But in actuality, "0" means anything in the 0 V to 0.8 V range, and "+5" means "greater than + 2.4 V". The ranges, then, are quite wide and analog voltages that fall within either range will be treated as if it were a real binary signal.

In Chapter 1, we discussed several different types of signal, of which digital is but one. You learned that analog signals can be continuously variable over a given range and domain. Such a signal can take on any value at any time. The sampled signal, on the other hand, can take on any value, but only at certain discrete times. Neither of these signals are compatible with digital circuits. The true digital signal of more than one bit can take on only values of the binary number system, and often is also time constricted.

Unfortunately, much electronic instrumentation produces either analog signals or sampled signals as the output. If these signals are to be processed in a digital computer, or using digital circuitry a *data converter* is required to change input values to a form that can be recognized by the digital circuitry; i.e., it must be converted to binary words.

There are actually two classes of data converters: *analog-to-digital converters* (ADC) and *digital-to-analog converters* (DAC). The ADC converts the analog voltage or current input to a binary word, as described above. The DAC performs the inverse process; it converts binary words to a proportional voltage or current output. The DAC may be used to drive analog display instruments, such as the oscilloscope or strip-chart recorder, or to drive analog control circuitry.

Transducers are devices that convert physical parameters such as force, pressure, velocity, position, temperature, light intensity, into an electrical signal. Most common transducers output a signal in the form of a voltage or current that is somehow proportional to the applied stimulus parameter. One common temperature transducer, for example, produces an output voltage proportional to the applied temperature. The scaling factor is given as 10 mV/°K. The freezing point of water is 273° K (0 °C). Note that the size of the Kelvin degree is the same as the Celsius degree, so convert °K to °C by subtracting 273). At freezing, then, the output voltage would be

$$273 \text{ °K} \times \frac{10 \text{ mV}}{\text{°K}} = 2.73 \text{ volts}$$

But before we can apply the data from a transducer or any electronic amplifiers and other circuits connected to the transducer, to a digital instrument, the output voltage (or current) must *first* be converted to a binary word. The data converter is used to accomplish this neat trick.

Similarly, some signals already exist in the form of voltages or currents. In medical electronics, for example, they use electrodes on the patient's body to pick up minute biopotentials (electrical potentials created by living cells). One commonly sought biopotential is the electrocardiogram, or ECG (also called EKG after the German spelling). This electrical signal is produced by the beating heart. Before any ECG monitoring or diagnostic computers can do their work, however, the analog biopotentials must be passed through a data converter to produce a series of binary words that represent signal.

Our last example, which will be the first class of data converters that we consider, is the DAC. There are two places where we would want to convert a digital signal—binary word—to an analog signal: display and control. In the case of the display, we might want to let human operators see the waveform on an oscilloscope screen, strip-chart recorder, X-Y recorder, or even an ordinary voltmeter, although there are non-analog methods the same job can be done. In the case of the control system, we might use an ADC to supply input data, and then output a computed result to a DAC to change or modify the system somehow.

Neither ADC nor DAC can represent all possible values or states of the analog signal. Remember, an analog signal can take on an infinite number of values within an assigned range (0 V to 10 V). The data converter, on the other hand, is limited to discrete amplitude values and discrete times. The former depends upon the range and bit-length (6-bits, 8-bits, etc.), while the latter depends upon the *conversion time,* or the time required to make a complete, valid, data conversion. Say we have an 8-bit ADC that is operated over an input voltage range of 0 to 10 volts. If eight bits are available, then there are 2^8, or 256, different binary states possible in the output word. All possible analog values must be shoehorned into 256 different states. The zero volts state is usually represented by the binary word 00000000. There are, then, 255 remaining combinations to represent all other analog values in the range. Clearly, then, only certain discrete input voltage valuescan be represented exactly. All others will contain a small error term. This leads us to the topic of converter *resolution.*

DATA CONVERTER RESOLUTION

The *resolution* of ADCs and DACs can be divided into three ranges: 2 to 6 bits, 6 to 10 bits, and over-10-bits. Those converters in the under 6-bits range are generally capable of accuracy in the 1 to 2 percent range, while those in the 6 to 10 bits range produce accuracies on the order of 0.1 to 1 percent. Even greater accuracy is possible in the over-10-bits class, but here some degree of caution is required. It is very easy to design a converter with 10 to 16 bit word lengths, but many have accuracies similar to 8-bit designs. Since we are always comparing an unknown analog voltage to a known reference source, the accuracy of the reference voltage (current) becomes important. The design of an 8-bit voltage reference supply is almost trivial, but the problem becomes more acute in over-10-bit designs.

Additionally, other problems tend to deteriorate the accuracy of the over-10 bit data converter. The gain error of any input amplifiers, for example, can easily become significant in this range. Another error is the hysteresis range of the voltage comparator circuit that is used to compare the reference and the input signal. The hysteresis is the minimum signal difference applied between the inputs of the comparator that will cause a change in the output state. This value ranged from 4 mV to 23 mV in a group of randomly selected LM311 comparators tested by the author.

Perhaps an example would best serve to illustrate this problem. In all ADCs, the noise should be less than one least-significant bit (1-LSB). Assume an ADC designed to operate over the 0 V to 10 V range of input signals. The minimum resolution is always given by:

$$E_{LSB} = \frac{10 \text{ volts}}{2^n}$$

where E_{LSB} is the voltage change required to change the LSB of the data converter, and n is the converter bit-length.
For an 8-bit ADC, then:

$$E_{LSB} = \frac{10 \text{ volts}}{2^8}$$

$$E_{LSB} = \frac{10 \text{ volts}}{256}$$

$$E_{LSB} = 0.04 \text{ volts} = 40 \text{ mV}$$

For a 12-bit ADC:

$$E_{LSB} = 10 \text{ V}/2^{12}$$

$$E_{LSB} = 10 \text{ V}/4096$$

$$E_{LSB} = 2 \text{ mV}$$

In the 8-bit case, the 40 mV LSB value is clearly greater than the 23 mV worst-case comparator hysteresis, so there is little cause for concern. But in the 12-bit case, the hysteresis of the comparator can be eleven times greater than the 1-LSB voltage. In this example, the noise bit is (LSB + 3), so our "12-bit" ADC will perform no better than an 8-bit ADC! It is generally conceded by ADC designers than words like "simple" and "low cost" form an exclusive set that does not include "high precision." One can, however, make use of low-cost or simple ADCs by properly recognizing true requirements. Over specifying ADCs can, indeed, be expensive.

When designers specify data converters they need to know the bandwidth of the input signal that will faithfully reproduce the waveform. All signals can be represented by a Fourier series of sine and cosine waves. Recall that a squarewave is made up from a fundamental and an "infinite" number of harmonics. The same is true of all other Waveforms: Only the pure sinewave contains no harmonics. The bandwidth required to faithfully reproduce or process an analog signal is approximately the frequency of the highest harmonic that can cause a recognizable change in the waveform if it were missing. For the human ECG described earlier, for example, it has been found that a bandwidth of 100 Hz is required. This means that amplifiers with a 100-Hz frequency response must be used to process the analog signal. It also sets the required *sampling rate* of the data converter. There is a theorem followed by circuit designers that demands a minimum sampling rate that is *twice* the highest frequency component present in the signal. Our 100-Hz ECG waveform, then, must be sampled at a rate of not less than 200 samples per second or there will be errors. In actual practice, faster rates are generally used. The data converter used, then, must be capable of making at least 200 conversions per second, or the waveform may be seriously compromised.

DAC CIRCUITS

It is generally best to consider the different types of DAC circuit prior to discussing ADC techniques. The reason for this seemingly illogical procedure is that DACs are used as an integral part of several popular ADC circuits.

The purpose of the DAC is to create an output voltage or current that is proportional to the product of a reference source and a binary input word. Considering only (for now) the voltage case:

$$E_o = \frac{A\, E_{ref}}{2^n}$$

Where E_o is the output voltage, E_{ref} is the reference voltage in the same units of E_o, A is the binary word applied to the DAC input (decimal expression), and n is the bit-length of the DAC input ($n = 8$ for 8-bit DACs).

Fig. 19-1. Binary-weighted resistor ladder DAC.

There are two principal forms of DAC circuit, and both are based upon simple resistor networks. One of these is the *binary weighted resistor ladder,* and the other is the *R-2R resistor ladder.*

Figure 19-1 shows the binary weighted resistor ladder. The ladder output is connected to the inverting input of the operational amplifier, which is also a summing junction. If you are unfamiliar with operational amplifier theory, let me recommend TAB Book No. 1550, Linear IC/OP-AMP Handbook-2nd Edition. The resistors in the ladder are said to be binary weighted, because their values are related to each other by powers of two. If the lowest value resistor in the circuit is taken as R, then the next higher value will be 2R, followed by 4R, 8R, 16R, 32R, etc., up to the nth resistor in the ladder which has a value of $(2^{n-1})R$.

In this circuit we show discrete switches used to turn on the resistors. These correspond to the digital inputs actually used in real DACs, which, incidentally, take on the form of electronic switches, so the analogy holds true. Switches B1 through B_n represent the binary word that is applied to the input of the DAC. Switches are used to connect each input to either ground (logical-0) or some reference voltage (logical-1). Switches B1 through B_n allow currents I1 through I_n to flow when they are closed, respectively.

Ohm's law tells us that each current is equal to the quotient of E and the value of the resistor. For example:

$$\begin{aligned}
I1 &= E/R \\
I2 &= E/2R \\
I3 &= E/4R \\
I4 &= E/8R \qquad I_n = E/(2^{n-1})R
\end{aligned}$$

The total current flowing into the summation junction of the op-amp is:

$$I_A = \sum_{i=1}^{n} \frac{A_i E}{2^{(i-1)R}}$$

where I_A is the current into the op amp summing junction from the resistor ladder network, E is the reference voltage in volts (V), and R is the resistance of R1 in ohms.

Note that the Σ symbol means "summation," and will be used elsewhere in this book. What it means, simply, is that we replace i with an integer that is one higher on each go-around, until n is reached. For example, consider a 3-bit converter ($n = 3$):

$$I_A = \sum_{i=1}^{3} \frac{A_i E}{2^{(1-1)}R} =$$

$$\frac{A_1 E}{2^{1-1}R} + \frac{A_2 E}{2^{2-1}R} + \frac{A_3 E}{2^{3-1}R}$$

$$I_A = \frac{E}{R} \left[\frac{A_1}{2^0} + \frac{A_2}{2^1} + \frac{A_3}{2^2} \right]$$

$$I_A = \frac{E}{R} \left[\frac{A_1}{1} + \frac{A_2}{2} + \frac{A_3}{4} \right]$$

where A_1, A_2, and A_3 are the bits at that input, and may be either 1 or 0. Note that in the previous equation, A_i will always be either 1 or 0; i.e., the switch is either turned on or turned off. Now, let's consider the operation of the circuit.

Operational amplifier theory tells us that

$$I_A = -I_f$$

and

$$E_o = I_f R_f$$

We can substitute the first equation into the second to obtain

$$E_o = -I_A R_f$$

We can now use this result to get:

$$E_o = -R_f \sum_{i=1}^{n} \frac{A_i E}{2^{(i-1)}R}$$

But E and R are constants, so we usually write them outside of the "summation" expression:

$$E_o = \frac{-ER_f}{R} \sum_{i=1}^{n} \frac{A_i}{2^{(i-1)}}$$

Example:

A four-bit DAC using the binary weighted resistor ladder technique uses a +2.56 V reference source. The value of R_f is 10 kohms, while R is 3.3K ohms. Find the output voltage for the binary input word 1101. (Hint: $A_1 = 1$, $A_2 = 1$, $A_3 = 0$, and $A_4 = 1$).

$$E_0 = \frac{-ER}{R} \sum_{i=1}^{4} \frac{A_i}{2^{(i-1)}} \qquad E_o = \frac{(-2.56)(10k)}{(3.3k)}$$

$$\left\{ \frac{A_1}{2^{(1-1)}} + \frac{A_2}{2^{(2-1)}} + \frac{A_3}{2^{(3-1)}} + \frac{A_4}{2^{(4-1)}} \right\}$$

$$E_o = (-7.76 \text{ V}) \left\{ \frac{1}{2^0} + \frac{1}{2^1} + \frac{0}{2^2} + \frac{1}{2^3} \right\}$$

339

$$E_o = (-7.76 \text{ V}) \left\{ \frac{1}{1} + \frac{1}{2} + \frac{0}{4} + \frac{1}{8} \right\}$$

$$E_o = (-7.76 \text{ V}) \quad 1 + 0.5 + 0 + 0.125$$

$$E_o = (-7.76 \text{ V}) \quad (1.625 = -12.61 \text{ V}$$

It is not always immediately apparent when examining ideal circuits presented in a textbook, but there are problems in trying to implement this circuit in actual practice. Bertrand Russell is attributed with the following quote: "There must be an ideal world, a sort of mathematician's paradise, *where everything works as it does in textbooks.*" The problem with this circuit is that the values of the input resistors tend to become very large and very small at the ends of the network. This becomes even more acute as the bit length of the converter becomes longer. There is little trouble when the bit length is less than eight-bits, but at longer lengths resistor values become ridiculous. Keep in mind that the highest resistor in an eight-bit converter will be 128 times the value of the lowest resistor. Yet the value of the lowest resistor must be such that E/R does not produce a large current that could burn out the resistor (or the op amp, for that matter). We also find that most easily obtained operational amplifiers cannot operate with the finds of input currents that will be found in the 128 R branch of the network. Consider the case above, where a 2.56 V reference source was used. If $R = 1K$ ohm, then 128R is 128,000 ohms. Current I8 will be $2.56/128000 = 0.000020$ amperes, or 20 μA. Most nonpremium grade operational amplifiers will not easily resolve this current. DACs built with these parameters might tend to "drop bits" in the least significant (I8) position. Additionally, have you ever tried to buy resistors in some of the odd "power of two" multiples of any reasonable R-value? They are not too easy to obtain on short notice and will cost a premium price when available. As a result, we rarely find the binary weighted ladder in real applications, and then only in 3 to 6 bit applications.

The R-2R ladder DAC circuit is considered superior to the weighted ladder in several respects. One is the fact that all of the resistors will have one of two values, R or 2R. This simplifies production of the unit, and results in a more consistent precision without a lot of trimming. Most commercial DACs use the R-2R resistor ladder instead of the weighted ladder.

Fig. 19-2. R-2R resistor ladder DAC.

The output voltage of the resistor ladder shown in Fig. 19-2 is applied to the noninverting input of an operational amplifier. In this case, we are using a high input impedance of the op amp to prevent loading of the network, and to provide buffering to the outside world. We could just as easily obtain some gain, but that is not too common in commercial IC DACs. Assuming that the value of R_L is very much higher than R or 2R, we can write the output voltage expression as:

$$E_0 = E \Sigma \frac{A_i}{2^1}$$

(The terms are defined similarly to the earlier expression given for the weighted ladder DAC).

Example:

A R-2R four-bit ladder DAC has a 2.56 V reference source and sees a binary word of 1101 at the input. Calculate the output voltage.

$$E_0 = (2.56 \text{ V}) \sum_{i=1}^{4} \frac{A_i}{2^i}$$

$$= (2.56 \text{ V})\frac{1}{2^1} + \frac{1}{2^2} + \frac{0}{2^3} + \frac{1}{2^4}$$

$$= (2.56 \text{ V})\frac{1}{2} + \frac{1}{4} + \frac{0}{8} + \frac{1}{16}$$

$$= (2.56 \text{ V}) (0.5 + 0.25 + 0 + 0.0625)$$

$$= (2.56 \text{ V}) (0.8125) = 2.08 \text{ V}$$

Full-scale output from the R-2R resistor ladder is given by the expression:

$$E_{o \text{ (max)}} = \frac{E (2^n - 1)}{2^n}$$

where $E_{o \text{ (max)}}$ is the maximum value of the output voltage E_o, E is the reference voltage, and n is the bit-length of the DAC digital input word.

Example:

Find the maximum output voltage of an eight-bit DAC that uses a 2.5 VDC reference supply.

$$E_{o \text{ (max)}} = \frac{(2.5 \text{ V}) (2^8 - 1)}{2^8}$$

$$= \frac{(2.5 \text{ V}) (256 - 1)}{(256)}$$

$$= \frac{(2.5 \text{ V}) (255)}{(256)}$$

$$= 2.49 \text{ V}$$

The input to a DAC is a binary word; it can exist only in those discrete states defined by the binary number system. The output of a DAC is dependent, in part, upon the digital input word. The DAC output voltage, therefore, can exist only with certain discrete values. Each successive binary number changes the output an amount equal to the change created by the least significant bit

342

(LSB). This voltage is often called the 1-LSB voltage, and is expressed by:

$$\Delta E_O = \frac{E}{2^n}$$

Consider an eight-bit DAC with a +10 VDC reference source:

$$\Delta E_o = \frac{(10 \text{ V})}{2^8}$$

$$= \frac{(10 \text{ V})}{(256)}$$

$$= 40 \text{ mV}$$

The 1-LSB voltage, ΔE_o, is the smallest change in output voltage that can occur. In most cases, if we let the zero output state be produced by the binary word 00000000, the maximum output voltage will be $E\text{-}\Delta E_o$.

ANALOG-TO-DIGITAL CONVERSION

Analog-to-digital conversion (ADC) is the inverse process to digital-to-analog conversion. The ADC takes an analog voltage or current input and produces a binary word at the output proportional to the analog input value.

Although many different analog-to-digital converter (ADC) schemes have been designed over the years, only a few have become available in the low-cost form of integrated monolithic or hybrid circuits. This chapter will consider only those that are readily available in that form. These include *single-slope integration, dual-slope integration, voltage-to-frequency conversion, ramp (or servo) flash,* and *successive approximation*. We will consider these techniques, and discuss their respective advantages or disadvantages.

Single-Slope Integration

The two integration techniques are examples of *indirect* analog-to-digital conversion. The analog signal is first converted to a voltage function of time, which is then converted to a binary, or BCD, number. In both designs, the principal components are an

Fig. 19-3. Single-slope integrator.

operational amplifier integrator, a comparator, gating logic, and a digital counter. The counter might have either binary or BCD outputs depending upon the application. In digital voltmeters (which ordinarily use dual-slope integration), for example, the counter would be BCD so that its output may be displayed on LED or liquid crystal decimal readouts. In a microcomputer system, on the other hand, either binary, or the special version of binary called twos' complement is used instead of BCD.

Figure 19-3 shows the circuit for a single-slope integrator ADC, while the timing diagram is shown in Fig. 19-4. During periods when

$$*E1 = \frac{1}{RC} \int_{T1}^{T3} E_{ref}\, dT$$

Fig. 19-4. Timing diagram.

344

no conversion is taking place, a switch keeps the integrator from accumulating a charge. But when a gate command is received, the integrator output voltage (E1) begins to ramp upwards, and the binary counter begins to accumulate clock pulses. When $E1 = E_x$, the comparator output drops LOW, and thus shuts off the flow of clock pulses into the counter. The output state of the counter at this time, then, is proportional to input voltage E_x.

Single-slope integration is easy to implement, but in its simpler forms suffers much from several inadequacies; the integrator output ramp may be nonlinear, or have the wrong gain (1/RC). This type of converter is sensitive to changes in clock frequency that occur during the conversion period. Although precision ADCs based on single-slope integration have been built, their added cost is sufficient to make other designs more attractive. The single-slope circuit is used primarily in very low cost digital voltmeters.

Dual-Slope Integration

Figure 19-5 shows the block diagram for a dual-slope integration ADC, while the timing diagram is shown in Fig. 19-6.

As in the case of the single-slope circuit, the principal components are an operational amplifier integrator, voltage comparator, and a digital counter. The comparator output will remain LOW if the integrator output is zero, and will be HIGH if there is more than a few millivolts of output from the integrator.

At the beginning of the conversion period, the control logic will momentarily short the integrator capacitor by closing switch S2. It will also insure that switch S1 is set to position A.

Fig. 19-5. Dual-slope integrator.

When switch S1 is in position "A," the integrator input is connected to the unknown input voltage source, and this causes the integrator output voltage to begin rising. This is shown at time t_{-0} in Fig. 19-6. As soon as E_A is more than a few millivolts greater than zero, the comparator output snaps HIGH. This will turn on the counter gate, allowing pulses to enter the counter circuit. The counter is allowed to overflow (in distinct contrast to the operation of the single-slope circuit!). The output-carry pulse from the counter tells the control logic section that the initial measurement is completed, and that S1 should be switched to position "B." The charge in the integrator capacitor at this time is a voltage function of time, and is proportional to the input voltage. With switch S1 in position "b," we find that the integrator input is now connected to a precision reference source. The polarity of the reference is such that it will tend to *discharge* the integrator capacitor. Since this reference is constant, the integrator capacitor will discharge at a constant, linear, rate. The counter has, by this time, passed through 0000 (on the carry-output it went to 0000) and has continued to accumulate pulses. It will continue to count until E_A is back down to zero. The value of E_A at time t_1 was proportional to E_x. At the same time, the counter display was zero (0000). Since the counter continues to increment as the integrator discharges, the counter state at the instant the gate is closed will be numerically the same as the voltage applied to the input.

The dual-slope integrator is a slow circuit, one of the slowest types of ADC circuit. But it offers several advantages that make

Fig. 19-6. Timing diagram.

Fig. 19-7. Servo, or ramp, A/D converter.

it useful in digital voltmeter circuits: relative immunity to noise pulses, relative immunity to clock frequency changes, and immunity to long term clock frequency drift. It's 10-40 mS conversion time is reasonable.

One major application for the dual-slope integrator is in digital voltmeters and multimeters (DVMs and DMMs). In that application, the slow conversion time of the ADC is not a disadvantage. In fact, it can be quite an advantage because it tends to average out certain types of noise. In many cases, we have a considerable amount of 60-Hz interference, but the dual-slope integrator DVM will average these to a very low value, provided that the conversion period is greater than 1/60 second, or approximately 17 ms.

Ramp or Servo ADC Circuits

The servo ADC is also known by the names ramp ADC and binary counter ADC. The latter name is derived from the fact that the circuit uses a DAC with its binary inputs driven by the outputs of a binary counter. Figures 19-7 and 19-8 shows the circuit to a servo ADC. The principal elements are DAC, counter, comparator, gate, and some required control logic. The reference potential source may or may not be present depending upon the nature of

347

Fig. 19-8. Typical timing diagram.

the DAC. The comparator is used to keep the gate open, allowing clock pulses to flow into the counter, as long as $E_o \neq E_{in}$.

A start *pulse* initiates the conversion cycle by causing the counter to reset to 00000000, and opening the gate to permit clock pulses to enter the counter. The counter will, therefore, begin to increment, thereby causing the DAC output voltage E_{-o} to begin rising. Figure 19-8 shows this action graphically. The DAC output will continue to ramp upwards until $E_o = E_{in}$. When this equality exists, then the output of the comparator drops LOW, turning off the gate. The binary number at the input of the DAC at this time is proportional to E_{in}.

The control logic section also senses the change in the comparator output level, and uses this indication to generate an output called the *end-of-conversion* (EOC) pulse. This pulse is critical to external circuitry, because there may be data on the output lines all of the time. The circuit cannot tell when the data is valid without an EOC pulse.

The conversion time of the servo ADC is variable, and depends upon two factors: the bit-length and the input voltage. While the bit-length of the DAC/ counter (which sets the bit-length of the converter) is fixed for any given circuit, the input voltage is not. The longest conversion time exists when the input voltage is full scale. The conversion time for full scale is 2^n clock pulses (where n is the bit-length). An eight-bit (n = 8 servo ADC, then, would require 2^8, or 256, clock pulses in order to make a full scale conversion.

348

Successive Approximation ADC Circuits

The successive approximation (SA) ADC circuit is somewhat more complex than the simple servo type, even though both are based on a DAC feedback loop. It is used in applications where the long conversion time of the servo ADC is not tolerable; the SA ADC is a lot faster for any given clock speed. For the SA circuit, we find that the conversion time is (n + 1) clock pulses. Consider a 10-bit ADC circuit in both servo and SA configurations. The values of the conversion times are:

Servo $T_c = 2^{10} = 1024$ clock pulses
SA $T_c = (10 + 1) = 11$ clock pulses.

From these, we can see that the SA ADC is 93 times faster than the servo for a full scale conversion. In applications where the conversion time is critical, the SA is clearly superior, despite its complexity.

How Does the SA Circuit Work?

The operation of the SA ADC is a little difficult to understand if we simply describe the electrical operation of the device. But with the use of the analogy of Fig. 19-9, we can more easily see the basic concept. This illustration shows a simple platform balance. The pointer will indicate balance (zero-center) only when the weight on the lefthand side is equal to the weight on the righthand side. In this example, let us stipulate that a weight W will deflect the scale exactly full-scale when pan No. 2 is totally empty. We have a set

Fig. 19-9. Analogy to the successive approximation A/D converter.

of calibrated test weights that have weights of W/2, W/4, W/8, W/16, etc (to whatever resolution, i.e., bit-length, that we require). If an unknown weight W_x is applied to pan No. 2, the scale will deflect to the right an amount proportional to W_x. To make our precise measurement, we start with W/2, and place it on pan No. 1. Three possible situations now exist:

- [ ] W/2 equals W_x, in which case the scale is zeroed.
- [ ] W/2 is greater than W_x (scale deflected to the left).
- [ ] W/2 is less than W_x (scale deflected to the right).

Of course, in situation No. 1, the measurement is completed, and no additional trials are necessary. But let us assume for the moment that one of the other situations exists. In case No. 2, W/2 is greater than the unknown weight, so we must remove W/2 and try again using W/4. This procedure is continued until we obtain situations No. 1 or 3.

In case No. 3, we find that W/2 was less than W_x. We now add more weights in descending succession of value until the scale balances, or the combined weights are greater than W_x. Of course, when the combined weights are equal to the unknown, then the measurement is over and the precise value known. But what about the situation where the value is between two of the possible combinations of weights? Remember that this is a digital circuit, and only certain discrete values of calibration weight are allowed. In this case, we have to set a protocol that determines which of

Fig. 19-10. Block diagram to the SA A/D converter.

Fig. 19-11. Typical operating sequence.

the values is most acceptable. The value of the last trial prior to the overweight trial is often selected for reasons of trade and the same is true of most SA ADC circuits.

Figure 19-10 shows the basic circuit of an SA ADC, while Fig. 19-11 shows a typical timing diagram for a hypothetical A/D conversion. In the SA type of ADC circuit, the scale is replaced with a specially constructed shift register, and the DAC represents the different values of weight applied to the "scale." The major components of the SA circuit are shift register. DAC, a set of output latches (one for each section of the shift register), and a control logic section. The output of the latches drive the inputs of the DAC,

A *start* pulse to the control logic section resets all latches to zero, and resets all shift register sections to zero. On the next clock pulse (T1 in Fig. 19-11), the output of the register is 10000000. This word places the output of the DAC at exactly one-half scale $(1/2\text{-}E_{fs})$. If the unknown input voltage is higher than this value, then the control logic section sets latch B1 HIGH. On the next clock pulse (T2), the next output register is set HIGH. The output of the DAC responds to the input word 11000000, and produces a $3/4\text{-}E_{fs}$ voltage. If, on any trial, it is found that E_{in} is less than E_o, the latch that is active for that trial is set LOW. But if the input voltage is more than E_o, then the latch is set HIGH.

Figure 19-11 is timing diagram of a three-bit successive approximation ADC making a conversion. Let us say that the full scale input potential is 1.0V and the unknown input voltage is 0.625 volts.

t1—The start pulse is received. Register B1 goes HIGH. The output word is now 100 and E_o = 0.5 volts. E_o is less than E_{in}, so the output latch for B1 is set for HIGH (i.e., logical 1).

t2—At the next clock pulse, register B2 is made HIGH. The output word is 110, and the output voltage E_o is 0.75 volts. Since E_{in} is less than E_o, the B2 latch must be reset to LOW. The output word at the end of this trial is 100.

t3—At the third clock pulse, register B3 is set HIGH and the output word is 101. The output voltage from the DAC (E_o) is 0.625 volts, so E_{in} = E_o. The B3 latch is set HIGH, so the output word remains 101.

t4—Overflow occurs. This tells the control logic that the conversion is over, and to a issue EOC pulse. In most cases, the overflow *is* the EOC, but in those that use a *status* output, the logic changes the level of the status bit.

A three-bit converter of our example required four clock pulses to complete the conversion. If a servo type had been used, the conversion would have been faster, but this will not happen once the bit length climbs to the five-bit, or higher, level.

At one time, the SA ADC was considered difficult to design and expensive to implement in any given piece of equipment. The problem was the amount of digital logic required to make the shift register, latches and the control section. But today, there are several integrated circuited successive approximation registers (SAR) on the market. A designer who wanted to start from "scratch," then does not need a lot of digital logic to make the circuit work.

Software ADC Techniques

Both servo and SA ADC circuits can be implemented in

software, with very little additional circuitry required. The critical logic is performed in software routines in a computer.

Figures 19-12 and 19-13 show how computer I/O ports can be used to implement a software/hardware A/D converter. One complete output port is used to drive the inputs of an eight-bit DAC. A single bit is used to monitor the status of the comparator output. A program must be written that will generate a ramp at the output port. This is done by creating a loop that will increment a binary word (usually held in a register) once for each pass through the loop. The elements required of the program are:

- ☐ Clear the accumulator (or other register used to hold the binary word).
- ☐ Increment the register by one.
- ☐ Output the binary word.
- ☐ Test the status bit at the input port.
- ☐ If the comparator loop is HIGH, then the program loops back to (2).
- ☐ If the comparator output is LOW, then the program must break out of the loop and use the accumulator contents as the data word.

Fig. 19-12. Using a minimum of hardware, a software binary ramp (servo) A/D converter can be implemented.

353

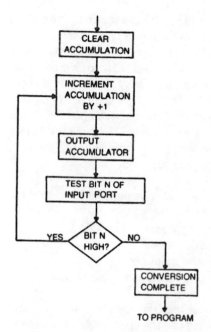

Fig. 19-13. Typical program flow chart for servo A/D converter in software.

A flow chart of this procedure is shown in Fig. 19-13. This particular program example will implement the servo type of A/D converter.

An example of a successive approximation A/D converter using the combined hardware/software approach is shown in Fig. 19-14. This particular circuit uses the 8080A microprocessor by Intel, and a Precision Monolithics, Inc., DAC-08 digital-to-analog converter. A slightly different circuit, plus program modifications, however, would allow the use of any of the standard microprocessor chips. This particular circuit is a memory-mapped A/D converter,and is turned on by bit A15 of the microprocessor. This means that the address assigned to the A/D converter must be in the upper 32K of memory (convenient because many microcomputers do not have more than 32K of memory. A sample program that will operate the A/D converter is shown in Table 19-1, while a flow chart for a typical three-bit A/D conversion is shown in Fig. 19-15. Also see Fig. 19-16.

Software implemented A/D conversion schemes take CPU time, so are not used unless there is plenty of CPU time available. Note that some uP chips, and many different DACs, are now so cheap that it is almost cheaper to dedicate a microprocessor chip, small-scale ROM, and a DAC to make a "discrete" A/D converter.

Fig. 19-14. Hardware for making an SA A/D converter in software with 8080 microprocessor.

355

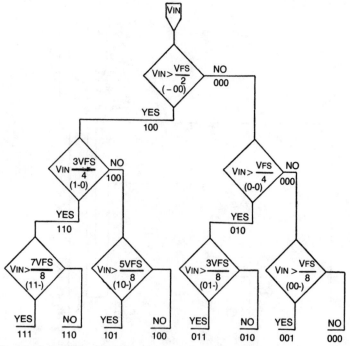

Fig. 19-15. Algorithm for making a three-bit SA A/D conversion.

Parallel (Flash) Converters

The parallel, or "flash," A/D converter is shown in Fig. 19-17. This converter is one of the fastest converter circuits known; its speed is limited only by the settling time of the voltage comparators used in the circuit. But this is not a free gift, because the output code is not one of the standard binary codes.

The flash converter circuit consists of $(2^n - 1)$ voltage comparators that are biased by a reference potential (E) through resistor voltage divider. Each comparator receives a bias that is 1-LSB different from the comparators above and below.

The very rapid conversion time is somewhat slowed by the need to decode the output prior to using it in an ordinary binary application. But the increased speed is so much that a few companies offer commercial flash converters in IC and hybrid forms.

Voltage-to-Frequency (V/F) Converters

A voltage-to-frequency converter is a type of A/D that converts

356

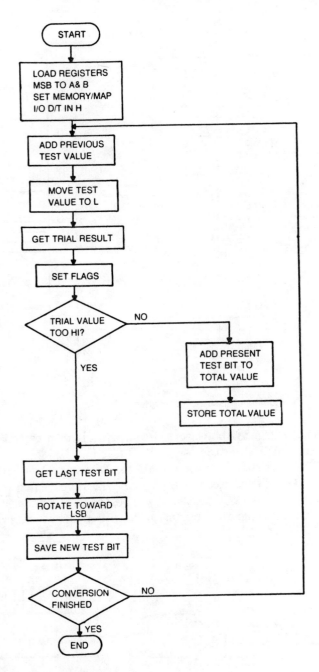

Fig. 19-16. Program flow chart.

Fig. 19-17. Parallel, or flash, A/D converter.

the input voltage to a proportional frequency. It is, then, essentially
a voltage-controlled-oscillator, or VCO (see Fig. 19-18).

There are two basic ways to use the VCO in A/D converter
service. One is to use a computer program to measure the frequency
of the output. The VCO signal is applied to one bit of the computer
input port. A computer program can be written that will measure

Fig. 19-18. V/F converter.

the *time* between successive input pulses, so that it can calculate the frequency; hence, the voltage. It is necessary only to know the scaling factor.

A second way to use the VFC is to gate its output into a binary counter. The gating pulse allows pulses from the VCO to input to the counter, and the counter output then becomes the binary word proportional to the input voltage.

VFC A/D converter are used where serial data transmission is mandatory,or where simplicity is dictated. But the technique is limited by dynamic range (although some companies offer VFCs with 1000:1 frequency change over the input voltage change).

Chapter 20

DC Power Supplies

For Digital Equipment

Digital electronics equipment often has much tighter power supply specifications than some other types of circuit. Regulation of the voltage to within very tight limits is required. We might also find things like *overvoltage protection, overcurrent limiting* (shutdown), *thermal protection,* and something called *safe operating area protection.* The latter, incidentally, refers to the collector power dissipation capability of the regulator transistor. In this chapter, we will discuss the basic principles underlying low-voltage DC power supplies, some different types of power supply circuits commonly used in digital equipment, and some problems/troubleshooting notes.

BASIC PRINCIPLES OF DC POWER SUPPLIES

The basic DC power supply consists of a transformer, rectifier, filter, and regulator. Other features are added to the supply for some special purposes, but these sections comprise the basic core for all digital power suppliers. In general, power supplies for digital equipment are low voltage designs, and may produce current levels that are quite substantial.

Transformers

The transformer (Fig. 20-1) reduces the 115-volt AC power mains potential *down* to a level compatible with the low voltage

Fig. 20-1. Transformer with voltages and currents defined.

requirements of the digital circuitry. Such transformers are called step-down transformers, and obey the following relationship:

$$E_p/E_s = I_s/I_p$$

Note from this equation that the current ratio is the inverse of the voltage ratio. A transformer that steps voltage down will show an apparently equal step-up of current. This concept can, unfortunately, become confusing to the new student who does not yet appreciate what is happening. In reality, secondary current I_s is determined only by secondary voltage E_s and the load resistance (R_L) connected across the secondary winding. The current step-up notion probably comes from the way this equation is often presented:

$$E_p\,I_p = E_s\,I_s$$

This equation tells us that I_p will vary as changing load conditions cause I_s to vary; as I_s goes up, I_p will also go up to keep the equation balanced.

Note that neither of the two forms this equation is presented have any loss terms. Real 60-Hz power and filament transformers are very efficient devices. True, there are losses involved, but most commercially available transformers are from 95 to 99.5 percent efficient, so the use of "lossless" equations is a justified simplification. The error terms are very small; so small as to be negligible.

Transformers are specified by their primary and secondary voltages, the secondary current, and the primary VA rating; i.e., the *volts times amperes*. Note that VA is the same as watts *only* in purely resistive circuits. In reactive circuits, an additional *power factor* is present, and it is equal to the cosine of the phase angle between I and E.

361

In most cases, the primary voltage rating will be 110 volts AC, 220 volts AC, or will be selectable between 110 and 220. In the latter case, the transformer will have two primary windings. They are wired in series for 220V applications, and in parallel for 110V applications.

The secondary voltage will be determined by the turns ratio of the transformer. Also, the secondary must be center tapped for some rectifier systems, but not if a *bridge rectifier* is used.

All voltages specified on transformers are rms (root mean square). But note that both peak (E_p) and peak-to-peak (E_{p-p}) become important in the design of DC power supplies. The relationships between E_{rms}, E_p and E_{p-p} for the sinewave are shown in Fig. 20-2.

The previous equations show that the primary VA and the secondary VA are equal at all times. But the primary VA rating is usually specified for the entire transformer. The VA rating is the maximum VA that can be obtained from the transformer without danger of overloading it. The reason for specifying the primary VA rating is that it is usually wound closest to the core, so cannot easily dissipate heat. As a result, the primary winding is usually easier to overheat. The primary VA rating must not be exceeded! While some manufacturers of equipment try to get away with it, this is not a smart practice. Failure to observe the primary VA rating may constitute a fire hazard and will almost certainly reduce the life of the equipment. Note that only transformers built to military specifications can normally be overloaded without danger.

FOR SINE WAVES:
$E_{rms} = 0.707 E_p$
$E_p = 1.41 E_{rms}$
$E_{p-o} = 2 \times E_p = 2.82 E_{rms}$

Fig. 20-2. Sinewave showing relationships.

Fig. 20-3. Rectifier forward biased allows current to flow at A, and rectifier reverse biased and no current flows at B.

Rectifiers

A rectifier (Fig. 20-3) is a device that will pass current only in one direction. This feature allows them to be used to convert bidirectional AC to unidirectional pulsating DC. Figure 20-3A shows a simple half-wave rectifier circuit. On the positive alternation of the AC cycle, the diode rectifier anode is positive with respect to the cathode; the diode is forward biased, so current will flow. But on negative alternations, the anode is negative with respect to the cathode, so the diode is reverse biased (Fig. 20-3B). No current will flow. The output waveform (Fig. 20-4) is unidirectional, but is not pure DC; it is a form of a DC called pulsating DC. Because of the missing half cycles, the average voltage at the output of a half-wave rectifier is only 0.45 times the applied rms voltage ($E_{1/2}$ = $0.45E_{rms}$). The missing half cycles also makes it necessary to have a transformer with a VA rating that is 40 percent higher than the required voltage times current.

The half-wave rectifier, then, is terribly inefficient because it does not use the entire AC input cycle. A rectifier that *does* use

MISSING
HALF CYCLES

Fig. 20-4. Output wave form of half-wave rectifier.

Fig. 20-5. Normal full-wave rectifier circuit.

the entire AC cycle is called a *full-wave rectifier*. Examples of the two common full-wave rectifier circuits are shown in Figs. 20-5 through 20-9.

Figure 20-5 shows a simple full-wave circuit using two rectifier diodes and a transformer that has a center-tapped secondary winding. The center-tap is designated as the zero reference point and is usually grounded. On any given half cycle, one end of the secondary will be positive with respect to the center tap, while the other end is negative with respect to the center tap. The situation reverses on every half cycle. We have designated the two ends of the secondary winding as points A and B for the sake of this discussion. One one-half of the cycle, point A is positive and point B is negative. On the other half of the cycle, point A is negative and point B is positive.

Consider the situation in which point A is positive and point B is negative. In this case, diode D1 is forward biased, and diode D2 is reverse biased. Current will flow from the center tap, through load resistor R_L, diode D1, and back to the transformer at point A.

On the second half of the AC waveform the situation is reversed; point A is negative and point B is positive. In this case, diode D1 is reverse biased and D2 is forward biased. Current will flow from the center tap, through the load resistor, diode D2 and

Fig. 20-6. Output waveform from a full-wave rectifier.

364

Fig. 20-7. Full-wave bridge rectifier.

back to the transformer at point B.

It is important to note that the current flows through the load resistor *in the same direction* on both halves of the AC cycle. This fact accounts for the double-humped waveform (Fig. 20-6) that is characteristic of full-wave rectifiers. The average output voltage from a full-wave rectifier is 0.90 times the applied rms voltage; exactly twice the output that was obtained from the half-wave circuit.

Note that the more efficient full-wave circuit is universally preferred in digital (and most other) electronic equipment. Only in transformerless designs, such as might be seen frequently in consumer electronic products, is the half-wave circuit preferred.

A full-wave *bridge rectifier* is shown in Fig. 20-7. This circuit does not need the transformer center tap, but this is only at the expense of two additional rectifier diodes.

As in the previous case, the two ends of the transformer are designated A and B. On one half of the input AC cycle point, A is positive with respect to point B. In this case, diodes D1 and D2 are forward biased, while D3 and D4 are reverse biased. Current

Fig. 20-8. Output waveform from a full-wave bridge rectifier.

365

Fig. 20-9. Circuit symbol sometimes used for bridge rectifiers.

flows from point B, through diode D2, load resistor R_L, and diode D1. It returns to the transformer at point A. On the second half of the AC cycle, the situation is reversed. Point A is now negative with respect to point B, so D3 and D4 are forward biased and D1/D2 are reverse biased. The current flow is from point A, through diode D3, load R_L, diode D4 and back to the transformer at point B.

Again, the critical factor for using the full AC sinewave is that the current flows through the load resistor in the same directions for both halves of the alternation. The average output voltage is the same as for the other full-wave circuit; i.e., 0.9 times the applied rms voltage. See Fig. 20-9.

The transformer used with the bridge circuit need not be center-tapped. The zero potential reference point is designated as the junction of the anodes of D2 and D3. This point is labeled negative (–), while the junction of D1 and D4 is labeled positive (+). Some prepackaged bridge rectifiers (see Fig. 20-9) have the DC terminals labeled with the " + " and " – " symbols, and the AC terminals labeled with either the letters "AC" or the sinewave symbol as shown.

One advantage of the bridge rectifier is that it will produce an output voltage that is twice that of the simple full-wave rectifier. This is due to the fact that the entire secondary of the transformer is used on each half cycle, instead of just half of the winding. But this is not a free gift, because we must still guard against exceeding the primary VA rating. The secondary current rating of most center-tapped transformers is assigned with the assumption that the simple full-wave circuit is being used. If a bridge circuit is used, instead, then only one-half of the rate current is actually available without exceeding the VA rating. The full rated secondary current can be obtained only if the transformer is designed specifically for use with bridge rectifiers.

Filters

Most electronic circuits cannot use pulsating DC as derived directly from the rectifier. They require instead pure (or nearly pure) DC. The pulsations are called *ripple*. Half-wave rectifiers produce a ripple component of 120 percent, at a frequency equal to the line frequency (60 Hz in the US). Full-wave rectifiers, on the other hand, produce a 48 percent ripple component at a ripple frequency of twice the line frequency. In the US, this means 60 Hz × 2, or 120 Hz.

A *filter* circuit smooths the pulsations to produce nearly pure DC. In the simplest case of the capacitor filter (Fig. 20-10), a single capacitor C1 is connected across the rectifier output, in parallel with the load. Although more complex filter circuits are known, most digital instruments use the simple circuit shown.

The action of the filter capacitor is shown graphically in Fig. 20-11. The heavy lines indicate the pulsating DC waveform without the filter (C1 were disconnected), while the dotted lines show the output with the filter capacitor connected. Capacitor C1 will charge to approximately E_p, but after the peak has passed, the charge will return to the circuit. The effect of returning the charge from C1 to the circuit is to fill in the area between the pulses as shaded the portion of the waveform. The filter reduces the ripple to a low percentage.

The value of C1 is not absolutely critical, only its *minimum* value is important. Higher values may be used with no deterioration of performance. In general, though, we require higher minimum values in half-wave circuits than in full-wave circuits. This is because of the lower ripple frequency in half-wave circuits. For most low voltage full-wave power supplies, the value of C1 will be 1,000

Fig. 20-10. Simple brute force capacitance filter.

Fig. 20-11. Output waveform with a capacitance filter.

μF/A or greater (some authorities prefer 2,000 μ/A). In really low-current applications (100 mA), however, this rule is modified to "no less than 250 μF, with 500 μF preferred". A 10-A DC power supply for a small microcomputer, then, should have a 10,000 μF filter capacitor. Some authorities, of course, would say 20,000 μF.

The circuit in Fig. 20-10 is the simplest form of DC power supply. But note that it is almost incapable of totally eliminating the ripple factor. Even if a very high value filter capacitor were used (and take a lot of space), the power supply would show some ripple in the output voltage.

A voltage regulator circuit will reduce the ripple factor almost to zero, even though the regulator's main function is to maintain a constant output voltage. One power supply manufacturer used to sell 6 to 16 volt bench power supplies with a very low ripple factor. The salesmen would tell you that it obtained this spec because the filter capacitor (15,000 μF) was "amplified" by a special electronic circuit to 1 farad. But in reality, the ripple reduction was caused by a voltage regulator circuit that reduced the ripple as much as would be obtained from a 1 farad capacitor. There was no "capacitance amplification", involved; only a 10 KμF capacitor and a voltage regulator!

Voltage Regulators

All DC power supplies have a certain amount of internal source resistance, R_s. When a load current is drawn from the power sup-

Fig. 20-12. Source resistance is a cause of regulation problems.

368

ply, a voltage drop, IR_s, will occur across the source resistance.

The actual value of the source resistance can be determined from Ohm's law. The voltage used in this calculation is the open-terminal power supply voltage obtained by disconnecting load resistor R_L and then measuring the output voltage from the supply when $I = 0$. The current used in the calculation is the short circuit current; i.e., the current that flows when the output terminals are shorted together. *Don't actually try to make this measurement!* Most DC supplies cannot long survive a direct output short without extensive damage. An alternative method for measuring R_s is this:

$$R_s = \frac{E - E_o}{I}$$

where R_s is the source resistance in ohms, E is the open-terminal output voltage, E_o is the output voltage existing under load R_L, and I is the current drawn when E_o is measured ($I = E_o/R_L$).

Voltage regulation is a measure of how stable the output voltage remains between load and no-load conditions. Figure 20-13 shows the mechanism that causes the output voltage change. Internal resistance R_s is effectively in series with load resistance R_L. The output voltage, then, can only be a fraction of the open-terminal voltage because R_s and R_L form a voltage divider. The output voltage at full load will be

$$E_o = \frac{ER_L}{R_s + R_L}$$

Voltage regulation is usually specified in terms of a percent-

Fig. 20-13. Equivalent circuit of a power supply.

369

age, which is calculated from the equation:

$$\%\text{Regulation} = \frac{(E - E_o) (100\%)}{E}$$

There are four basic forms of voltage regulator circuit: *Zener diode*, *Zener-referenced series-pass*, *feedback*, and *switching*. Others are often merely different forms of these basic types.

Zener diodes have a property that allows them to be used as voltage regulators. Figure 20-14 shows the curve for a typical Zener diode. In the $+E$ region, the diode is forward biased and will behave exactly like another PN junction diode. But in the reverse region ($-E$), the diode is reverse biased. The Zener diode is a little different from ordinary silicon diodes in this region. From zero volts, to a point marked V_z, the diode *is* like other diodes. No current, or only a tiny leakage current, flows. But at V_z the diode breaks down and conducts a reverse current. The value of V_z tends to remain constant for a wide range of applied voltage and reverse current level. It is to V_z that the Zener will regulate the output voltage.

Figure 20-15 shows a typical low-current regulator based on the Zener diode. Diode D1 is connected in parallel across the load. Resistor R1 used to limit the current to a level that is safe for the Zener diode. Without R1 the diode would burn out.

Fig. 20-14. Zener diode characteristic.

Fig. 20-15. Zener diode voltage regulator.

The simple Zener voltage regulator circuit is used only for light duty work. A general rule-of-thumb is that the load current must be held to 10 to 20 percent of the Zener current; hence, the limitation to low current applications.

A better solution is shown in Fig. 20-16. Here the Zener diode is used as a low-current reference voltage source in a series-pass circuit. In this circuit, transistor Q1 is the actual control device, and D1 is a reference. The output voltage is given approximately by:

$$E_o = V_z - V_{be}$$

The maximum output current is approximately the normal Zener load current (I_b in this case) multiplied by the beta of transistor Q1. This assumes, of course, that neither the maximum collector current, nor the maximum collector dissipation ratings are exceeded.

Another form of series pass voltage regulator is shown in Fig.

Fig. 20-16. Series-pass voltage regulator circuit.

Fig. 20-17. Feedback amplifier voltage regulator.

20-17. This circuit is called a feedback voltage regulator and depends upon comparing the output voltage E_o with a reference potential E_{ref}. In this circuit, the bias applies to series-pass transistor Q1 is determined in part by the output of unity gain differential amplifier A1. Amplifier A1 is often called an *error amplifier*. It is a differential amplifier because its output is proportional to $E_{ref} - E_o'$. If voltage E_o changes from the value set by E_{ref}, as when a load current change occurs, then the sample E_o' applied to the error amplifier also changes. The amplifier output will respond by changing an amount and direction that tends to cancel the error.

A *switching regulator* is shown in simplified form in Fig. 20-18. In this circuit, a switching transistor (shown as an SPDT switch in Fig. 20-18) is used. The transistor must have a very low value

Fig. 20-18. Switching regulator.

372

of $V_{ce(sat)}$ for the circuit to operate properly. The switching regulator operates at a high frequency, such as 15 kHz. The LC network forms a low-pass filter to reduce the high-frequency component. Coil L must have a very high Q. If the transistor switch is turned on, then the full input voltage is applied to the output. But if the transistor is turned off, then zero volts is applied to the output. The circuit works by pulsing the switch on and off for periods that reflect the output voltage. In some circuits, the duty cycle is adjusted so that a 50 percent on-time results in the proper output voltage. The on-time of the switch is determined by the output of a pulse width modulator, which is in turn controlled by the output of the error amplifier. If the output voltage drops, then the PWM increases the on time of the switch, causing more energy to flow into the low pass filter. But if the output voltage rises, then the on-time of the switch decreases.

In general, switching regulators (of which, Fig. 20-18 is but one type) are considerably more efficient than series pass circuits. Keep in mind that the switching regulator may, at first glance, look much like a series-pass circuit because Q1 is in series with the V_{in} supply. But the tell-tale sign is the use of a low-pass LC filter at the output and a pulse-width modulator.

Actual IC and Hybrid Regulator Devices

The easiest way to obtain low-voltage DC power supplies of 1 ampere (and possibly to 5 or 10 amperes!) is to use a fixed output voltage three-terminal IC regulator. These devices, which are housed in transistor packages, can be obtained in standard voltage level values up to 24 VDC.

Three-terminal IC voltage regulators are available in several different, but similar, families. Three general catagories are available: 100 mA, 750-1000 mA, and 1000-1500 mA. There are also several individual types available in current ranges up to 10 amperes. Figure 20-19A shows several of the common case styles used for voltage regulators. Note that they are all transistor cases. The letters, which are used in the type number designation for the regulator, tell us the case style and the approximate current limitations. Consider the LM-309 device for example. There are at least two different LM-309s on the market: LM-309H and LM-309K. Both are +5 VDC regulators. The LM-309H, which is housed in a TO-5 transistor package, is a 100 mA device. The LM-309K, on the other hand, is in a TO-3 package so can dissipate more power. The rating of the LM-309K is 1 ampere in free air.

Fig. 20-19. Three-terminal regulator case styles are similar to transistor cases at A, and typical circuit for positive regulators at B.

Up to 1.5 amperes (some manufacturers claim 2A) can be drawn if the device is properly heat-sinked. The T-package, incidentally, is the transistor TO-220 plastic package (sometimes called P-66), and will dissipate up to 750 mA in free air and 1 ampere if properly heat-sinked.

The most common three-terminal IC devices are the LM-340 and 78xx series regulators.For LM340 devices, the voltage rating is given by a suffix separated from the main number by a hyphen, while the current rating is given by the case-style letter:

$$LM\text{-}340 \ (x) \ - \ (yy)$$

The case-style letter is placed where (x) is shown, while the voltage rating is placed where we show (yy). Only the letters H, K, or T are used. The LM-340K-12, then, is a 12V positive voltage regulator in a TO-3 (i.e., K-style) package. Because of the package style, it is capable of passing up to 1 ampere. The LM-340T-15, on the other hand, is a 15V positive regulator in a T-style (750 mA) package.

The 78xx series replaces the "xx" with the voltage rating. A 7805, then, is a 5 volt positive regulator, while the 7812 is a 12V positive regulator. It is not usual to include the case style as part of the type number in this series of devices. Note, also, that

Motorola oftentimes uses their own case style (the one with the hole in the middle) for the plastic package devices. These may, however, bear the 78xx designation.

Negative regulators are also available in the same package styles and voltage ratings. These are given the designations 79xx and LM-320. The same protocols are followed for the type number. An LM-320K-15, for example, is a 15-volt, negative voltage regulator in a TO-3 (1 ampere) package. Note that there is a pinout difference between positive and negative voltage regulators! They are *not* pin-for-pin compatible. If you try to connect a negative voltage regulator into a circuit where the positive regulator had been used before, it is simply not enough to reverse the polarity of the input in order to obtain the required negative voltage output. You must also swap the correct pins on the IC regulator.

The basic circuit for positive voltage regulators is shown in Fig. 20-19B. Transformer T1, diodes D1 and D2, and filter capacitor C1 are the same as in any unregulated DC power supply. Note also, that a bridge rectifier may be substituted for D1/D2; in fact, it is in most cases. Capacitor C1 is the filter capacitor, and should not have a value less than 2000 μF/A. The function of capacitors C2 and C3 is to improve the immunity of the circuit to noise impulses. Capacitor C4 is optional, but is used to improve the transient response of the output circuit. The value of C4 should be approximately 100 μF/ampere.

Until relatively recent times there were no voltage regulators on the market that were capable of delivering more than 1 ampere (without loads of heatsinking!). But now, there are several on the market. One of the first was the Fairchild Semiconductor LM-323 (TO-3 package) that was capable of delivering 3 amperes. Fairchild and Motorola now offer several models in the 5 ampere range, and at least one type is available that will produce 10 amperes. Lambda Electronics (515 Broad Hollow Road, Melville, NY 11746) produces a series of three-terminal IC regulators in the TO-3 package. They offer 1.5 ampere, 3 ampere, 5 ampere, and 8 ampere models. These are the LAS-15xx, LAS-14xx, and LAS-19xx series, respectively. As in the 78xx series, the voltage number replaces "xx" in the type number. An LAS-1905, then, is a 5 volt regulator able to deliver up to 5 amperes.

An exception to the Lambda numbering system is the LAS-19CB, which delivers 13.8 VDC at 5 amperes. This device is designed mostly for "12 volt" DC power supplies for operating CB rigs from the AC power mains. Note that most "12 volt" automobile electrical systems are actually 13.8 to 14.4 volt systems.

All three-terminal regulators have a certain minimum and maximum input-output differential voltage ($E_{in} - E_o$) rating. A typical number is 2.5 volts, although some ask for as much as 3 volts, or as little as 2 volts differential. The maximum voltage is related to the amount of power that can be dissipated safely in the collector of the internal series-pass transistor. For most devices, the maximum input voltage will be in the 30V to 40V range. The minimum I/O differential is the smallest differential that will allow the regulator to operate properly, while the maximum voltage is the level that may cause damage to the device if exceeded. Note that some hybrid regulators that deliver very high currents (i.e., more than 15 amperes) will not deliver full output current if the voltage is operated near the maximum. These devices will have a dissipation rating in watts that must not be exceeded. We must, therefore, derate the current specification if the device is operated near the maximum allowed input voltage. Consider a 5-volt, 5-ampere regulator, for example. It may have a 50-watt dissipation rating, and a 35-volt input voltage. The power actually dissipated is:

$$P_d = I(E_{in} - E_o)$$
$$= I(E_{in} - 5)$$

So, if the maximum input voltage is the allowed 35 volts,

$$P_d = (5A)(35 - 5)$$
$$= (5A)(30) = 150 \text{ watts!}$$

From the above it is clear that we cannot expect to draw the full rated output current if we operate the device near the full allowed input voltage. We can find the maximum allowed output current under those conditions by rearranging the equation to solve for I when E_{in} is 35 volts (or whatever voltage is actually used) and P_d is 50 watts. Similarly, we can also find the maximum allowed input voltage that will permit us to draw the full 5 amperes by rearranging the equation to solve for E_{in} when I is 5 amperes and P_d is 50 watts. In the case shown, the full rated output current is available only for input voltages less than 15VDC. The same reasoning also applies to other types of voltage regulator.

Lambda Electronics also offers a line of four-terminal IC regulators. These are similar to three-terminal devices, but have one additional terminal that permits us to adjust the output voltage to any point within a given range. An example of this circuit is shown in Fig. 20-20. The device is housed in a four terminal version of the TO-3 transistor case. A simplified internal circuit diagram is shown in Fig. 20-20A. Note that it is a feedback type of regulator, and has thermal, safe operating area, and overload protection. The latter becomes very important to the survival of

Fig. 20-20. Lambda four-terminal regulator internal circuit block diagram at A, and case pinouts at B, and a typical circuit at C.

377

the power supply if a catastrophic short circuit develops. The series-pass element inside of this regulator is a Darlington pair of transistors. This allows us to take advantage of the extremely high beta possible in the Darlington configuration ($H_{fe(Q1)} \times H_{fe(Q2)}$).

The case style for this regulator and the pinouts are shown in Fig. 20-20B. Since the case is similar to TO-3, we find that TO-3 heatsinks will work nicely with the regulator, provided that the extra holes are drilled to accommodate the other two terminals.

The Lambda four-terminal regulators can be set to any voltage that is within their range using potentiometer R_p (see Fig. 20-20C). The range for most models of this series is about 4 to 30 volts. The LAS-15U, LAS-14U, and LAS-19U produces output currents of 1.5, 3, and 5 amperes, respectively. With the exception of the control pin and the potentiometer, the four-terminal regulator uses the same general circuit as the three-terminal devices.

High current power supplies of over 5 to 10 amperes are a little tricky to design properly. You can't expect to just slap a Zener diode into the base circuit of a high gain transistor and make it work properly. We suddenly run into thermal problems and certain other difficulties. Additionally, it is too easy to buy ready-built, hybrid, voltage regulators in current levels up to around 30 amperes. The Lambda models in this range are:

LAS2xxx	5 amperes	85 watts
LAS3xxx	10 amperes	140 watts
LAS4xxx	10 amperes	170 watts
LAS5xxx	20 amperes	270 watts
LAS7xxx	30 amperes	400 watts

The LAS5205 is a 20 amperes, 5-volt regulator that will dissipate over 200 watts. Similarly, the LAS7215 produces 15 volts at almost 30 amperes.

Figure 20-21 shows an example of a circuit using a Lambda hybrid voltage regulator. The value of the voltage adjustment trimmer potentiometer should be *at least* ($0.25E_o \times 1000$ ohms/volt).

The exceptions to this rule are the 5-volt models, in which the minimum value of the resistor is 3000 ohms. In actual practice, the minimum value for any given specimen of voltage regulator may be a lot higher. In a power supply which I built for my microcomputer, the actual value required was in the 11,000-14,000 ohm range.

Fig. 20-21. Circuit for a Lambda hybrid regulator circuit.

Note that both pins 1 and 20 are positive-input-voltage terminals. This can lead you astray in some cases because they have different input-output voltage requirements. Pin No. 1 will operate with a +2.5 volt I-O differential, while pin No. 20 requires a +7.5 minimum differential! Pin No. 1 is the normal high-current input, while pin No. 20 powers the internal control circuits. If a device uses a common +8-volt, unregulated high-current power source for the regulator, then another low current supply must be designed and built for pin No. 20. Fortunately, the current requirements here are low, and most microcomputers have a +12 to +15 VDC supply which can be used for pin No. 20.

OVERVOLTAGE PROTECTION

Most integrated circuits must operate within certain voltage limits. While the CMOS devices have a relatively wide operating voltage range, the TTL device is very limited. In no instance must the voltage rise above 5.6 volts DC, or damage to the device will result. And it become damaged in a relatively short period of time! Most digital power supplies have +8 volts on the input side of the regulator, and it is a high current source. If the series pass element in the regulator shorts c-e, then this +8 volts is applied to the +5-volt line and will probably damage most of the TTL devices in the circuit. To prevent this catastrophe, use an overvoltage protection circuit at the output of the regulator. Although Lambda makes a two-terminal module for this purpose, most will use a discrete circuit such as shown in Fig. 20-22. This circuit is called an *SCR crowbar* because it is a brute force approach to solving the

379

Fig. 20-22. SCR crowbar overvoltage protection circuit.

problem. High-current SCR D2 is connected in parallel across the output of the supply. It is normally turned off. Diode D1 is a +5.6-volt Zener diode. If the line voltage increases above +5.6 volts, then D1 conducts and thereby turns on the gate of the SCR. This action causes the SCR to turn on and short the output of the power supply. In most cases, there will be a series fuse or circuit breaker. In other cases, the primary fuse protecting the transformer will blow or the SRR will just simply sit there and absorb the excess current flow.

OUTPUT CURRENT LIMITING

A short circuit in the digital device can cause immense damage due to the high current nature of the power supply. A short on one of the PC boards in an unprotected piece of equipment can burn up all of the printed circuit lands in the +5-volt and ground lines all the way back to the power supply. In addition to that, it can destroy the power supply in quick order. The answer is output current limiting. A simple example of an output current limited circuit is shown in Fig. 20-23. Transistor Q1 is the regular series-pass element for the voltage regulator. Transistor Q2, however, is used as a shutdown switch for the current limiter. The bias for Q2 is obtained across a small series resistor that has a value of

$$R = 0.6/I_{max}$$

Fig. 20-23. Output current limiting circuit.

When the maximum current is exceeded, the base-emitter voltage needed to turn on Q2 is exceeded. This causes the collector-emitter path of Q2 to drop in resistance, effectively shorting together the base and emitter of Q1. A small increase over maximum output current causes "soft limiting", but as the current increases even more the "hardness" of the limiting increases rapidly until total shut off is reached.

TYPICAL POWER-SUPPLY CIRCUITS

Figures 20-24 and 20-25 show two typical power supplies that might be found in digital equipment. The circuit shown in Fig. 20-24 is a ± 12-volt, low-current (1-ampere) power supply. It is based on the LM-340 positive and LM-320 negative voltage regulator ICs. Note that both of these circuits are identical to circuits presented earlier in this chapter, except for the arrangement of the transformer secondary and rectifier. At first glance, you see a bridge rectifier, but note that the transformer is center-tapped! This is not a contradiction, because one-half of the secondary and one-half of the bridge rectifier are used for each side of the power supply. The

Fig. 20-24. Dual-polarity power supply.

381

Fig. 20-25. High-output current, low-voltage power supply with sense line.

382

bridge is not full-wave, but is called a two-diode bridge, or half-wave bridge, rectifier.

The circuit in Fig. 20-25 is a 10-amperes, + 5-volt DC source for the TTL circuitry in a microcomputer. The rectifier (B1) is a 25-ampere type built in modulator form. The filter capacitor (C1) has a rating of 18,000 μF, but in many cases expect to find values up to 100,000 μF. In a version of this circuit that I built for my microcomputer an 80,000 μF unit was used. Transistor Q1 is the series-pass element, while Q2 is the current limiter. The heart of the circuit is a Motorola IC voltage regulator. This circuit device can be used alone at currents up to 500 mA, but needs a series-pass element at higher output currents. It is housed in a 10-pin transistor-like package that has physical dimensions similar to the small TO-66 power transistor case.

Note the *remote sense* terminals in this supply. This is very common on the + 5-volt supply of digital equipment. It seems that, at the current levels expected, ordinary ohmic losses in the conductors between the output of the regular and the actual circuit boards can be quite high. They may look small on first glance. But when you consider the fact that many TTL devices begin to act up if the operating voltage drops below + 4.75 volts, then it becomes apparent that millivolts are important. The sense line is to the error amplifier in the internal feedback regulator (inside the Motorola IC). The sense line is connected to the point on the PC board where we want to insure a + 5-volt operating potential. In actual practice, this line is placed either at the farthest extreme, near the most critical board, or about 3/4 the way between the supply and the farthest PC Board.

TYPICAL POWER-SUPPLY PROBLEMS

One service shop that kept records of the actual faults found in warranty equipment discovered that more than 60 percent of the total number of service jobs were either in the power supply itself, or generated symptoms that affected the power supply voltages. While this information will not surprise old timers in the electronic service business, it is good advice to the newcomer: Look at the power supply voltages *first*. Regardless of the actual numbers and percentages on any specific equipment, it is true that you can solve a lot of problems quickly by taking a brief look at the DC power supply, especially when symptoms appear weird.

Chapter 21
Test Equipment
For Digital Servicing

One of the first questions asked by anyone responsible for an electronic servicing business is: "What kind of test equipment is needed, and how much does it cost?" When color television first came on the scene, we learned about things like color-bar generators, vector scopes, and wideband oscilloscopes. But even then, the standard of the industry was recurrent sweep models for servicers. Triggered sweep oscilloscopes were regarded as too expensive, exotic, complicated, and all of above. But today, the typical TV servicer will use a recently designed triggered sweep oscilloscope routinely.

If you are already in the electronic servicing business, then you may own all of the test equipment that you need for some types of digital servicing. Indeed, only a small additional investment need be made, and that in small items.

The ordinary test equipment common to all electronic servicing will usually fill the bill in simple digital servicing. You will, for example, need a good portable voltmeter or multimeter. Most digital servicers prefer the digital multimeter types of instruments. If you have an "old fashioned" analog meter, however, don't rush out and buy a brand new digital job unless you want to, or need an additional meter. That old VOM/VTVTM will work just as well for the rough measurements needed in digital work. Keep in mind, though, that digital servicing tends to be field servicing. Some of the new digital multimeters are more suited to field work than most analog meters. They are thinner and cost less in battery power. Those that use CMOS ICs and a liquid crystal display will last a long time on a battery change—and will be easier to maintain.

If you have any bench power supplies, then they will be useful in digital electronics servicing providing that they are regulated. An old car radio "battery eliminator" probably will not do the trick, however. Bench supplies that do not meet the specs can be upgraded by addition of a couple three-terminal voltage regulators (see Chapter 20). It is especially necessary to provide one or more sources of +5 VDC at currents up to 5 amperes. One service shop used a Lambda LAS-1905 (5 VDC at 5 ampere) three-terminal regulator inside of an old Delco 12-volt battery eliminator. A new heatsink on the back panel and a pair of heavy duty five-way (banana) binding posts did the trick. The input to the LAS-1905 was merely the main power from the Delco P1200, set to +8 volts (or more).

People entering the digital electronics service business cold might wish to contact the test instrument manufacturers for a package deal on needed instruments. Tektronix, Inc. (POB 500, Beaverton, Ore., 97005) offers their TM-500 line of low-cost modular test instruments. The TM-500 instruments are designed for field service, as well as bench service, work. They are plug-in modules and can be configured in any of several mainframes.

If there ever occurs a situation where I am told that only one type of test equipment or instrument will be allowed for me to perform a troubleshooting job, then my answer will be the *oscilloscope*. This is the one instrument that is truly versatile and almost indispensable. Color TV servicers long ago learned that they needed to update their old-fashioned "audio" range oscilloscopes that had served so well in B&W receive troubleshooting. They bought 5 and even 10 MHz models for color service. In fact, those test equipment makers that specialize in equipment for TV servicers were prone to call these "wideband" scopes *Color TV oscilloscopes*.

In digital service work, the specs on that old 'scope will probably take another leap upwards. The vertical sensitivity is not too awfully critical, but that bandwidth may be important. The 4.5 MHz *color TV* scope might work out for much service work if clock speeds in the digital circuit are low and rise time of the scope is not critical. The rise time is related to, but not identical with, vertical frequency response. In general, on a properly designed oscilloscope, the rise time will be found approximately from

$$T_{(sec)} = 0.35/F_{(hz)}$$

where T is the rise time of the vertical amplifier in seconds, and F is the vertical amplifier frequency response in hertz.

As the frequency of the clock or the rise time of the pulses in a digital circuit become higher, the requirements placed on the oscilloscope used by the servicer becomes more critical. If you do not own a good oscilloscope, then plan to buy one. Both Tektronix and Hewlett-Packard will be glad to send you their catalog.

While some cheaper versions of scopes seem to have similar specs, they simply do not perform as well as the ones that cost more. A good scope is a tool of the trade.

Not all features of all models will be needed in any given shop, so one should examine the catalogues, and request that a sales person bring one out to the shop for an "audition" under field conditions. Some will even leave a demonstration unit with you for a week or so, if you seem like a good prospect.

Keep in mind the need for portability in your oscilloscope. If most of the work will be permanent on-site assignments, then a largish mainframe unit could be justified. But if all of the work will be temporary service call assignments, please consider the poor slob who has to carry that darn oscilloscope four city blocks from the nearest parking space to the site! An extra heavy oscilloscope may tend to be left in the car more than not, with resulting backtracking lost time when the servicer finds the scope *is* needed.

LOGIC PROBES

After making an impassioned plea for you to rush right out to buy a multi-kilobuck oscilloscope, let me tell you a little story. One hospital where I worked had a complicated pulmonary test instrument that was made of mostly digital circuits. This instrument was a massive collection of digital circuits in a six-foot-high 19-inch rack panel. The field engineer *owned* a high quality Tektronix portable oscilloscope. But in two years, I never saw the scope anywhere but in the trunk of his car! When he came to service the equipment, he brought a briefcase tool box and two test instruments: a digital multimeter and a *logic probe*.

You ask just *what* the blue blazes is a logic probe? That's a fair question. It is a pencil-like probe (actually most look like a fat scope probe) that detects logic levels and clock pulses. Some will allow you to generate a pulse or a logic level, while others merely detect and indicate activity in the circuit. Some will have three lights as indicators: HIGH, LOW and clocking. The clock detector is capacitor coupled so mere levels will not fire the lamp.

The reason why the field engineer for the pulmonary test equipment never needed a scope is that most of the circuit actions, except the clock, took place at slow speeds, or were merely levels that would come up or down as switches were manually closed. Most of the time, only the *existence* of a level or pulse needed to be confirmed. For that special case, a logic probe was superior to an oscilloscope. Such probes cost $38 to $295 and represent a low investment for a large return.

LOGIC ANALYZERS

In digital servicing, *timing* is often the keyword in finding a problem. This means, *at least*, a dual-trace oscilloscope in most cases. Recently, though, the advent of multi-bit data paths, microcomputers, etc., have made it mandatory that we be able to examine the states of several bits *simultaneously*. If the events are periodic—repetitive, then an oscilloscope with a multi-channel electronically switched input circuit might be useful. But it is often the case that some event will trigger the circuit, and the events of interest to the servicer will be temporary. They will look like mere transients to the oscilloscope. A storage oscilloscope could trap some types, but they are very expensive as frequency increases. Another answer is to use a logic analyzer. This instrument will capture and display on a CRT the states of the logic bits under examination. Then the servicer can analyze almost at leisure events that took place in nanoseconds. This type of instrument is used in design labs and those service jobs involving complex instruments.

Some logic analyzers will display the levels in the fashion of a waveform timing diagram. But in computer work it is often the case that we want to see the binary words that passed along a data or address bus during the event. Some instruments will display as characters on a CRT screen, the binary words that were received. Some will allow a single bit to trigger the analyzer, and will then display, in either hexadecimal or binary form, the 256 words prior to and after the word existing when the trigger was received. The trigger word will be accentuated by excessive brightening of its representative on the CRT.

Another form of logic analysis involves signature patterns. This technique is similar to vectorscope troubleshooting in color TV. But in the case of signature analysis, the instrument displays a pattern that indicates the performance properties of the circuit in response to a standardization stimulus—a certain pattern of operation within the circuit, such as a standard diagnostic program run by a computer being serviced. This technique becomes important when dealing with large-scale digital circuits, where it is difficult to analyze in the old stage-by-stage manner that worked so well in AM broadcast receivers of three decades ago.

Most logic analyzers are expensive instruments. For the majority of digital service jobs, they will prove unnecessary. But given the cost, it would be wise to weigh this factor carefully when taking on jobs that will require the instruments. The question is, as always, would it be cheaper given the incidence of service to buy a logic analyzer, or to be a board swapper and let the factory buy the logic analyzer (*not* an unaccepted practice in field service)?

TOOLS

If you intend to go into field service of classy digital equipment, then learn who Jensen Tools and certain of their competitors are. They are a sales organization that makes up field service tool kits in brief case tool boxes. This turns out to be both practical and a very necessary image builder. The tools themselves are pretty much the same sort of thing that will be carried by a TV servicer in the tool/tube/module caddy. But image is very important when you are charging a lot for going out and fixing the equipment. How well you are regarded depends upon how the customer perceives you— and that affects the willingness to pay the freight! If you show up in dungarees and carrying a tool caddy, you are then a workman to be disdained. But if you show up in a suit carrying the same darn tools in a tool brief case, you are seen as a field engineer worthy of your hire.

Besides, I have found that those brief cases are a lot easier to lug around than a bulky caddy, and their size limits the amount of trash that you carry into the customer's site. It is often too easy to go back for special tools than to carry everything everywhere.

Chapter 22

Some Common Problems

There are two principal areas where digital equipment can have recurrent problems: power supply and temperature. These two considerations cause many different types of problems. Many of the problems are either recurrent, toughdogs, or both.

Any piece of digital equipment that contains substantial amounts of TTL devices will generate a lot of heat. The high current requirements of *any* fast (high-frequency) logic family will mean a large amount of heat energy being dissipated. Unfortunately, many devices, especially when used in combination with (or synchronized to) other circuits, become flakey when the operating temperature goes above a certain limit. As a consequence, one may pay close attention to overheating problems. One stand-alone minicomputer, for instance, uses forced air blowers to keep the temperature down. The manufacturer has placed a hole in the cabinet to be used as a temperature test point. The service technician inserts a regular clinical medical fever thermometer into the hole, and allows it to come to equilibrium. The machine is considered to be operating within the temperature specification if the temperature at the test point does not exceed 102° F, or 38.9°C.

Power supplies frequently encountered in digital equipment, especially larger systems, are usually equipped with thermal shutdown circuits. These circuits will turn the power supply off when the over-temperature conditions exist. Also, series-pass and switching transistors tend to short out *collector-to-emitter* at temperatures over some point in the 125 to 200° C range. They are derated at temperatures above 30° C.

Temperature, therefore, can easily be a critical parameter in digital equipment. One might suppose that manufacturers would see to these matters in the original design of the beast. And usually they do! But, sometimes, Murphy's law takes over and some machine is designed with a deficiency in the design of the thermal characteristics; in other words, there was a screw up somewhere back in the engineering department. We also find instances where a customer will operate the equipment greater than originally specified by the manufacturer. Large, and even medium size, computers, for instance, are usually housed in a well-air-conditioned room. Some of those rooms get so cold that the operators must wear sweaters everywhere but in the immediate vicinity of the machine. If an instrument, machine, or computer is designed for a, say, 75° F room, and is actually operated at a much higher temperature (say 90° F), then the heat transfer away from the sensitive circuits is reduced. This will cause overheating and possible malfunctions.

The solution is to either reduce the ambient temperatures markedly, or increase the heat flow away from the circuits. In some cases, both strategies are needed. In some cases, the customer might be willing to air condition the room. But if this is too expensive, impossible, or impolitic, then the other alternative must be examined.

In most cases, electronic equipment must be cooled with forced air. Ventilation holes must be placed so that air blown across the affected circuitry can escape and carry the heat with it.

There are right and wrong ways to eliminate heat from equipment. Figure 22-1 shows several possible situations. In Fig. 22-1A a finned heat-sink, as might be found dissipating the heat generated by the series-pass element of the DC power supply (itself a major producer of heat) sensitivity function. The heat-sink should be physically placed so that air from the forced air blower travels between the fins. If it travels perpendicular to the fins, then it will not see maximum exposed surface area. Blowing air across the maximum surface area tends to maximize the heat transfer.

A similar situation exists with respect to printed circuit boards inside of a cabinet. The fan or blower should be positioned to blow across the component side of the printed circuit boards. Figure 22-1B shows the incorrect way to position the fan, while Fig. 22-1C shows the correct placement. Note that in Fig. 22-1C, the fan should be placed to blow evenly across all of the boards. If one of the boards

Fig. 22-1. Correct and incorrect airflow patterns across heatsink at A, incorrect airflow through stack of PC boards at B, and correct airflow at C.

is known to run hotter than the others, however, then position the fan to blow predominantly over that board. It might also be required to add an *additional* fan or blower to cover that particular board. There are also times when it is advisable to *increase the rating* of the existing fan. In one case, I replaced a 56 cu. ft./min. (7-watt) whisper fan with a 115 cu. ft./min. (14-watt) fan of the same series. This cured a difficult thermal problem for my employer.

Keep in mind that *heat rises*. Therefore, fans and blowers should be placed in the lower parts of the cabinet, and vent holes located near the top. Figures 22-2 and 22-3 show several situations involving equipment cabinets. In Fig. 22-2A, you see a small cabinet situation, such as might be found in desk-top, or bench top, instruments. The fan is mounted on the rear panel, and vent holes are placed along the top edge of the cabinet. Forced air circulates from the fan, upward through the printed circuit boards, and out of the cabinet via vent holes. Forty to 60 cu. ft./min. is *usually* sufficient in small equipment cabinets, unless some element is present that would add to the heating normally associated with TTL devices.

Fig. 22-2. Correct airflow through PC card cage when side mounted at A, and bottom mounted fan at B.

Figure 22-2B shows a similar situation in which the fan is bottom mounted. Air forced up through vents at the top of the cabinet. Some enclosures have a vent port, or series of holes, along the top edge of the rear panel. Others will use a perforated screen top-cover to allow heat to escape. The latter, incidentally, is not limited to cabinets with bottom mounted fans.

We often see equipment installed in standard 19-inch wide relay racks. Most of the principle manufacturers of relay racks also make dual squirrel cage blowers to cool the equipment. The blower is usually mounted at the bottom rear of the cabinet, and (sometimes) hoses, that resemble automobile radiator hoses, are used to distribute the forced air to the modules installed in the cabinet. A large exhaust port along the top edge (front or rear) will permit the heated air to escape. It is usually the practice to mount the worst heat generating modules closest to the top of the cabinet. Otherwise, rising heat from lower units tends to overheat the upper units.

Heat generated in any type of electronic equipment can be substantial. I can remember a 5000-watt (output) AM radio station that was wasting almost 6000 watts of anode heat dissipated by their water-cooled final amplifier and modulator tubes. They were able to reroute the heated water from the inside of the transmitter to external radiators, and it was sufficient to heat the transmitter building and the engineering department section of the main building. All winter long, even on the coldest days, it was usually too hot inside of the area by "waste" heat from the electronic equipment.

In another case, I can remember a medical computerized ICU/CCU patient monitoring system that averaged one expensive service call every 10 days or two weeks. The failure rate dropped to one very *three months* after blower fans were installed in the cabinet and ventilation holes were added.

POWER LINE PROBLEMS

Problems with AC power mains can affect almost any type of electronic equipment. My window air conditioner, for example, bears a label that in effect says the unit may burn out if operated during power company brownouts and line noise problems.

Brownouts occur when the AC power mains voltage drops from its specified range of 105 to 120 volts AC (rms) to 90 to 105 volts. Under these conditions, the DC voltage available from the rectifiers also drops. If it drops below the level required at the input of the DC voltage regulator circuits, then the voltage applied to the electronic circuits will be interrupted, even though there is DC applied to the regulator input.

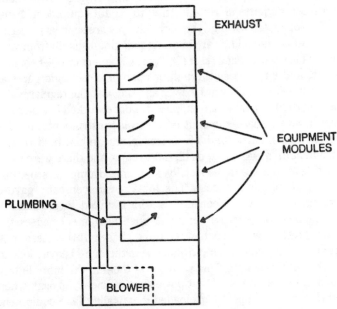

Fig. 22-3. Airflow distribution in rack mounted system.

It is never good to lose the DC supply voltage in any electronic equipment. In a digital computer, however, the problem is compounded by the fact that you may lose the data that is stored in volatile (semiconductor) memory. There are some circuits available that will guard against this problem, such as uninterruptible power supplies and special circuits that will dump memory contents to a disc when a power failure occurs. These usually operate from batteries on board the computer, or from the energy stored in the DC power-supply filter capacitors. Regardless, it is a sad state of affairs for the user when the machine goes down.

Another significant problem in AC power mains-operated equipment is that there is a considerable amount of noise present on the line. An oscilloscope analysis of many power lines will show frequent high-voltage transients and often a constant level of nonimpulse, high-frequency noise. Those nice AC sinewaves shown in our elementary electronics textbooks (including some by this author) actually belie the real situation. The truth is quite different. There are, according to one study, at least 10 transients on the average *residential* power line every day. Industrial and commercial lines can be expected to be even worse because of the types of equipment that may be operated on those lines.

The problems presented by having a trashy AC power mains source can be particularly upsetting to digital equipment. Noise spikes propagated through the power supply are seen as pulses by the digital circuits. They are in effect indistinguishable from valid signals. They will set and reset flip-flops, increment or reset counter circuits, and either pass through or open or close logic gates at inopportune times. They are viewed as being just like regular, valid, pulses normally seen in the equipment. I once worked in a building that contained a large amount of electronic instrumentation that was used in medical and life sciences research. Much of this instrumentation was based on digital circuits, and there were more than a few computers, especially after microcomputers became easily available at low cost. The local power company gave a premium rate to those consumers who installed load-balancing equipment at the service entrance. Such equipment senses the current level drawn on each of the three phases. It then uses triac switches to disconnect some circuits from the more heavily loaded branches of the building and reconnects them to the more lightly loaded branches. It seems that power line efficiency is peak when all three available phases are loaded identically. That equipment represents a significant savings to the building management and the local power company. But it also throws high-voltage spikes down the lines inside the building, and those spikes play heck with electronic equipment. One humorous aspect of the situation occurred early in the building's history, and amply illustrates the problems: All 125 freshman medical students flunked an examination that had been graded by computer. The exam was multiple choice, and the students marked IBM answer sheets. The answer sheets were then read by an optical scanner, which was connected by an interface cable to an IBM keypunch machine. These machines, working together, would read the answer sheets and then punch out one IBM Hollerith card for each student. These cards were then hand-fed into a card reader connected to the university's IBM 370 at the computer center across campus. The computer would grade the exams, rate the student's performance against others taking the examination, and do certain statistical chores for the professor. But in one case, the optical scanner was reading the answer sheets during a time when the power switches were sending transients down the line. The keypunch saw them as valid signals, and punched a lot of extra holes in the student's cards. You can imagine the uproar when 125 of the nation's future physicians saw the posted computer printout of their scores!

Power line conditioning, then, can become critical when digital equipment is used. Some cases of intermittent problems are so frequently due to power line problems, that the suppression practices that are next discussed are almost standard fare with digital equipment field service personnel.

In ideal situations, one would suppose that transients should be suppressed *at the source*. If the cause is known to be a certain motor, arcing contacts, or some similar source, then we can apply a capacitor, rf shielding, or an rf suppression filter, as needed.

But the situation facing most digital service personnel is a little different; the ideal is seldom realized. We usually find that the source is unknown, too multitudinous, or is impossible to suppress. We also find situations where it is merely uneconomical to suppress the noise at the source. It might be a lot cheaper to suppress the noise at the service entrance of the digital computer. In those cases, we will have to *condition* the AC line and the DC power supply.

One quick fix for many power source woes involves using a line voltage regulator/isolation transformer (Fig. 22-4) between the power mains and the equipment. In fact, some manufacturers either build such a transformer into their equipment, or include it among the site preparation specifications given to the customer.

Fig. 22-4. Use of isolation transformer.

In the latter case, they may decline to honor their warranty if the transformer is not included. These transformers are models designed specifically to attenuate pulse transients on the line, and provide some regulation of the AC voltage supplied to the equipment connected to the secondary. Both *Sola* and *Topaz* make various models for applications ranging from small desk top computers to large mainframe computer systems.

Another solution, one used by many manufacturers, is to provide rf filters at the point where the AC power enters the equipment. Such a filter circuit is shown in Fig. 22-5. The rf chokes are selected to have a negligible reactance at 60 Hz, but considerable reactance at radio frequencies. The capacitor values are selected to have a low rf reactance and a very high 60-Hz reactance—usually 500 pF to 0.1 μF is specified. The entire filter assembly is mounted inside of a shielded enclosure, and will be located very near the service entrance to the equipment. Some manufacturers use chassis mounted AC power receptacles that have built-in, shielded rf filters. This means that no unfiltered AC power wiring will be inside the equipment cabinet.

Fig. 22-5. Brute force rf filter.

396

Fig. 22-6. Use of MOV and Zener diodes to suppress transients.

Two other approaches are shown in Fig. 22-6. The device shunting the primary winding of the transformer is a *metal oxide varistor* (MOV). General Electric is a prime manufacturer of these devices. The MOV is, in essence, a voltage sensitive resistor. If the voltage across the MOV remains at normal line values (170 volts peak), then nothing happens. The MOV has a high impedance under these conditions. But if a large transient comes along, the peak voltage will exceed the threshold, causing the MOV to exhibit a large drop in impedance. Because of the very short duration of the transient, the sharp drop in resistance across the AC line does not result in an appreciable increase in the total current drain. It will not blow the fuse.

The back-to-back Zener diodes across the secondary are used to clip transients. They will have a breakover voltage that is greater than normal peak AC values encountered in the circuit.

OTHER GLITCHES

A *glitch* can be defined as an unwanted pulse in a digital circuit. Glitches are caused by different things, but they all raise hob with circuits. They will do all of the nasty things attributed earlier to power line noise pulses, but are sometimes a lot harder to nail down.

Figure 22-7 shows one cause of glitches. Pulses A and B are supposed to be coincident at the trailing edge. In situations where this coincidence is important, the actual time difference is not zero, but some small difference t (trace C). This can be propagated as a glitch, an unwanted and possibly harmful pulse.

Fig. 22-7. Generation of a glitch.

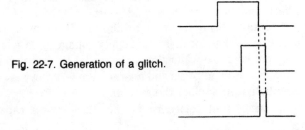

397

The situation described above can be caused by any of several factors. One of these is the propagation delay of the digital IC devices being used. They normally require a few nanoseconds in order to pass an input change over to the output, and this causes a slight delay. If a particular IC or one particular manufacturer's IC is a little too slow or a little too fast, then problems occur. This is especially common when devices of two different manufacturers or two different logic families are mixed. But note that it can occur anytime. The solution sometimes is to change ICs until one is found that works. But sometimes the problem is due to a difference in a particular series IC, or those made by a specific vendor. In that case changing vendors, lot numbers, or series will help. The 74H, 74S and 74LS devices with similar suffix numbers are pin-for-pin compatible, yet have different operating spaces and propagation times.

In still other cases, where the problem seems to be related to timing, the equipment manufacture will instruct the field service people to lengthen or shorten a pulse duration. On several occasions, for example, I have seen manufacturers change the RC time constants of one-shot multivibrators in order to eliminate a glitch problem.

Another type of glitch is caused by failure to properly bypass the DC power supply lines on the printed circuit boards. Analog electronic PC boards can often get away with very few bypass capacitors; i.e., some in the power supply and a couple on each PC card. But digital circuitry is a lot more critical in this respect. All of the fast, high-current drain, IC logic families are subject to this type of problem. TTL seems especially susceptible. The rapid pulse-like turn-on times seen in TTL circuits cause abrupt changes in the DC current drawn from the +5 VDC power supply. Fast rise time leading edges have a very high frequency component, so the distributed inductance and resistance of the +5 VDC/ground traces on the P.C. board become important. When the TTL output goes LOW, an instantaneous increase in the current drain occurs. If the power supply were perfect, of course, then this would not have any affect and no problem would exist. But because the supply distribution wiring looks more like an RL network (Fig. 22-8), expect a sudden, local drop in supply voltage. The solution to the problem is to use wider (low-inductance) PC board lands for the +5 V and ground paths and a large number of bypass capacitors distributed throughout the system.

Various protocols exist for how the bypass capacitors are

Fig. 22-8. Equivalent circuit for TTL power distribution.

distributed. Almost all of them want to see a 1 to 25 μF tantalum capacitor at the point where +5 VDC enters the PC board (usually right at the card-edge connector). Some also want to see a 25 to 250 μF/tantalum capacitor at the farthest extreme of the +5 VDC line, and a capacitor value of 100 μF/A at the output of the voltage regulator. The latter improves the transient response of the regulator, while the others improve the transient response of the PC board DC distribution system.

Almost all designers like to see an ample supply of small disc ceramic capacitors distributed all across the printed circuit board. These are of critical importance in many circuits. Some designers specify 0.001 μF/disc at the +5 VDC terminal of every TTL IC used in the circuit. Others use 0.001 to 0.1 μF every two or three TTL ICs. Regardless of which is preferred, some form of loyal bypassing must be supplied. Current stored in the capacitors can be dumped into the circuit when the protected IC suddenly increases its current demand (usually when the output goes LOW). This prevents the glitch and effectively cancels the effect of the distributed resistance and inductance of the line. Good bypassing will filter out most of the glitches that are created and propagated by the DC power system on any given PC board.

BUSES

Most digital equipment uses data or address buses. These are parallel signal lines used to form a digital word in parallel format. In a typical eight-bit microcomputer, for example, they will have an eight-bit *data bus* and usually a 16-bit address bus. For our example, let us consider the address bus. Most of these machines can supply up to 65-536 (2^{16}) different eight-bit memory locations. This is due to the bit-length of the address bus; 16-bits can produce up to 65,536 different, unique, binary word combinations. In order to make a 16-bit bus, however, we need 16 parallel conductors!

The idea of a bus came into being to reduce the amount of wiring and circuitry needed in digital equipments. We can parallel a large amount of circuitry on a common bus. Decoders at each device, or circuit, are then used to allow response only to a unique code. Consider eight-bit memory locations normally used in digital computers. The eight-bit data bus is connected in parallel to all 65,536 different memory locations (or some submultiple, as is common when 1K and up memory chips are used). Similarly, the 16-bit address bus is connected in parallel to all 65,535 sets of address lines. Any given memory location, however, will be effectively turned off unless it sees the unique code for its address on the address bus.

But buses can cause problems. No device has a perfect input, especially TTL devices. Every device that you connect to the line draws a small amount of DC current, even when it is officially turned off. When dozens (or dozens or dozens) of inputs are connected in parallel with a bus line, the voltage supposedly present in the HIGH condition is dragged down by ordinary voltage divider action. In some cases, the voltage reduction is so severe as to place the potential of the HIGH state in a gray region where TTL devices have difficulty determining HIGH from LOW.

One cure for this problem is to use *passive termination* of the bus lines (Fig. 22-9). We terminate each of the lines with a pull-up resistor to the + 5 VDC source. When the line is LOW, then the bus end of the termination resistor is brought to ground and does not interfere. But when the line is HIGH, it sees a high impedance to ground and that makes the voltage on the bus wire closer to + 5 VDC.

An *active termination* is essentially the same idea, but uses a small, independent, regulated power supply to accomplish the same mission. There is some indication that active termination is superior.

Ringing is another problem on buses, especially long buses. The bus signal lines are usually long (thereby increasing R and L), and quite close together (increasing C). Applying a pulse, then, is the same as ringing an equivalent RLC circuit. Following each pulse we see a damped sinewave oscillation. Termination, especially the

+5 VDC

(ALL 3300Ω)

Fig. 22-9. Use of pull-up resistors in a passive termination of the data bus.

active variety, decreases the duration of the oscillation to almost nothing.

RFI FROM DIGITAL EQUIPMENT

Digital circuits operate on pulses, so any digital equipment will usually have a considerable amount of pulse energy present all of the time. This is a situation that is ripe for the generation of radio frequency interference, or RFI. You can see an example of this phenomenon by holding a pocket calculator, especially a model using LED readouts, close to an AM radio receiver. You will hear a large amount of hash-like interference. My first microcomputer had a "demonstration" program that used RFI to play *The Star Spangled Banner* on a nearby AM radio receiver, while printing an asterisk and "x" pattern on the video terminal screen that resembled the US flag.

But putting games aside, RFI can be a serious problem. Digital equipment must often coexist with other equipment. Sensitive radio

401

receivers, or RFI prone scientific instruments simply cannot operate properly with so much RFI trash floating around.

Part of the problem is radiation via the AC power line wiring. Fortunately, the same procedures that protected the digital equipment from trash already on the power line will also prevent transmission of the RFI from inside of the equipment. This is especially true of rf filters and isolation transformers.

But also be concerned with radiation from other than the power line. A prime source of radiation is interconnection cables, such as between the keyboard and the computer (although any interconnection cable can be the culprit). A 3 to 5 foot cable to a peripheral device is a very sufficient transmitting antenna.

Radiation also comes from the circuitry inside of the cabinet. This is a phenomenon long familiar to communications people. Shielding of the offending circuitry is often the best solution for both types of radiation. Any book on RFI problems should be more than sufficient to provide troubleshooting ideas. Some of the best, incidentally, are found in the amateur radio sections of book catalogs.

UNINTERRUPTIBLE POWER SYSTEMS

An uninterruptible power system, or UPS, is often specified for computers that *must* operate all of the time. Any computer that contains critical, real-time data is a candidate for such a system. As the name implies, a UPS cannot allow the power system to be interrupted. There are several ways to accomplish this feat. One is to use batteries to power the computer. The AC power system is used to charge the batteries. If loss of the AC power occurs, a control circuit turns on an emergency generator system to continue charging the batteries. The batteries will contain sufficient energy to operate the computer for the short period of time when the gasoline, or diesel, engine driving the generator is starting up.

Another UPS system is shown in Fig. 22-10. This technique also uses an engine-driven generator, but the generator is always used to operate the computer. Normally, the engine is off, and an AC motor turns the AC generator shaft. The output of the AC generator is then used to power the computer. A line voltage sense circuit determines when the AC line power is interrupted, and turns on an engine-starting circuit. When the engine comes to full speed, it will be capable of turning the AC generator shaft by itself.But what about the time between loss of the mains power and full-speed operation of the engine? This is taken care of by a *flywheel!* The

Fig. 22-10. One type of uninterruptible power system.

flywheel is normally kept turning by the AC motor. By its nature, a flywheel does not come to a grinding halt when driver power is lost. There is enough energy stored in the flywheel to keep it running for a long time. This gives the engine sufficient time to start.

Chapter 23

Other Equipment

Many years ago, when I was still fresh from a high school electronics shop course, I fancied myself as a hot-shot radio/TV repairer. One of the first jobs given to me when I reported to my first *real* job was to—yeeucck—clean and lube a record player. The point is that servicers of exotic electronics equipment often must wear multiple hats. They often service a range of products that include electrical, mechanical, and electromechanical devices. Digital servicers have the same problem, but some of their electromechanical junk seems a little more exotic than on old RCA 45RPM phonograph; in fact, the equipment can be more complicated than the computer or device it serves.

PRINTERS

Various types of printers are used to provide hard copy output for computers and other digital instruments. Some printers are little more than a drop-in OEM module that has been placed in a side corner of the front panel of the main instrument. Other printers are teletypewriters (Teletype, IBM Selectric, etc.), while still others are mechanical wonders that print whole lines at a time. The actual printer used is determined by the application and selections of the designer. There are, however, only a few technologies used. These can be broken into two basic categories: *impact* and *nonimpact*.

Impact printers, much like an ordinary typewriter, use a hammer or something similar to strike a printhead of some sort.

Nonimpact printers use thermal or electrical techniques to deposit the letters or numbers on the paper.

One of the simplest printers is the numerical printer of Fig. 23-1. This type of printer is frequently seen in digital instruments that output a number or value. It is based on the simple mechanical counter (Veedor-Root mechanism). Such a counter is advanced one place for each pulse applied to its electromagnet. Note that these are not electronic counters, but electromechanical. They are like the little hand-operated counter you may have seen. Each time a button is pressed, the counter increments by one. In this case, however, the counter "button" is the armature of a relay or solenoid. Each time the relay is energized, the counter advances by one. The counter operates by having the digital instrument output the number of pulses that corresponds to the desired count. The counter will advance to that point, at which time a set of solenoid operated strike hammers impact on the counter wheels. A piece of specially treated paper is placed between the hammers and the print head, so the letters are left on the paper. This technique is used only in the simplest, lowest-cost printers. They are quite popular, however, because many instruments only need to output a numerical value. In many cases, the printer will print many lines because it will advance the paper after each print and make a fresh impression using new data.

One of the earliest types of printers used with computers was the old-fashioned Teletype® machine. Although there have been several models over the years since World War II, the one that has proven most popular with computer manufacturers is probably the Model 33. This machine is able to respond to ASCII code, instead of Baudot code, and is therefore more compatible with computer standards. The Model 33 uses a print mechanism such as shown in Fig. 23-2. The characters are embossed on a metal cylinder that

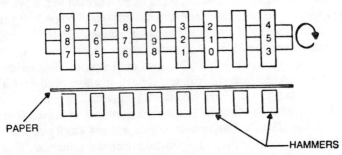

Fig. 23-1. Numerical printer using a mechanical counter system.

405

RIBBON
Fig. 23-2. Print cylinder such as used in teletypewriters.

is mounted upright with respect to the paper. The mechanism senses the ASCII code and mechanically decodes it. This operation causes the cylinder to rotate to the position containing that letter, and then the hammer strikes the cylinder. A ribbon is placed between the cylinder and paper, so an impression is left on the paper. The teletypewriter operation is more like that of an ordinary typewriter in that respect.

Very popular for many years has been the IBM *Selectric®* typewriter mechanism. This mechanism is the so-called "golf ball," or "flying ball" system. The balls are interchangeable for different type fonts. The Selectric is an office typewriter, but for years IBM offered a Selectric I/O model for computers. Even today, while IBM does not make an I/O model anymore, a number of vendors buy Selectric mechanisms from IBM and build them into their own cases. The secondary manufacturer will perform the modifications to make the Selectric computer-compatible and then sell it to the user. An example of these are the Andersen-Jacobsen terminals.

Another printer mechanism is the daisy wheel of Fig. 23-3. In this type of machine, the characters are embossed on pads at the ends of the spokes of a wheel. The wheel rotates constantly, and when the desired character is opposite the print space, the hammer strikes the pad, making the impression on the paper. Neither the cylinder (Teletype), flying ball, nor the daisy wheel need special paper, as they use inked ribbon to make the impression.

One other type of impact printer is the dot-matrix impact print head of Fig. 23-4. In this system, the characters are formed in a 5×7 dot matrix by little pins emanating from the print head. The decoder circuitry will cause the correct pins needed for the character

406

Fig. 23-3. Daisy wheel printhead.

to stick out of the head. These pins impact the paper and leave the impression. Either ribbon or specially treated paper can be used.

Some high-speed printers use a belt system, such as shown in Fig. 23-5. The characters are embossed on a belt that is constantly rotated. When the desired character is opposite of the print space,

Fig. 23-4. Dot matrix impact printhead.

Fig. 23-5. Print band system.

a hammer strikes and leaves the impression.

Most thermal and electrical printers use the 5×7 dot matrix system, but the dots are formed in a manner different from the impact version. In the thermal type, each dot is a hairpin loop of wire that is heated to incandescence when the dot is needed to form a character. In the electrical type, the dot forms one terminal of an electrical circuit. The platen of the printer forms the other contact. When a current passes through the wire, the connection between print and platen is made through electrically conductive special paper. A black dot is left at the point of contact. Both thermal and electrical printers are very popular; we will, however, see far more of the thermal types.

PAPER TAPE READERS

Paper tape was once used extensively in teletypewriter work to transmit repeated messages or to transmit high-speed messages. In the latter case, several operators would cut the message onto tape. Then a high-speed teletypewriter would transmit it at a rate faster than could reasonably be followed by a human operator. Paper tape was also used extensively as a means for program storage, especially in the early days of computer technology. It is still used widely, however, even though it has been largely eclipsed by magnetic tape and magnetic discs. Most tape machines will both read and punch tapes.

Figures 23-6 and 23-7 show two methods by which paper tape can be read. The older method used sense wires or pins to make electrical contact with a grounded plate underneath the tape. If the bit at that location was a mark (a1), then there would be a hole in

Fig. 23-6. Electrical sense wires for reading paper tape and punch card.

the tape at that location and the sensor would make contact. But if the bit was a space, then there would be no hole so the sensor would not make contact with the plate because paper is an insulator.

In more modern equipment, however, an optical method does the same job. An array of photocells or phototransistors takes the

Fig. 23-7. Optical reader does the same job as Fig. 23-6 mechanism.

place of the grounded plate, while light-emitting diodes (LEDs) take the place of the sense wires. If there is a hole in the paper (MARK), then light from the LED shines on the photocell. But if there is no hole in the paper, light cannot reach the paper. The optical method is also used in nonpunched paper tape in which inked dots are used to represent the bits. Card readers use essentially the same technology to read the data punched into the card.

MAGNETIC STORAGE DEVICES

All computers have a limited on-board or internal memory. In older computers, this memory was magnetic cores, so the on-board memory was called *core*. Today, solid-state memory is used. The total internal memory capability of a computer is determined not by the *physical* size of the memory (solid-state is generally smaller than core), but by the length of the address bus. In a typical microcomputer, the address bus is 16 bits in length, so the computer can support up to 2^{16}, or 65,536, different memory locations. Many applications require far more memory, however, instead of building monster computers with megabit address buses, though other forms of temporary or semipermanent storage are used. The on-board memory can be loaded with blocks of data from the external memory as needed. Alternatively, the computer can write data from the internal memory to the external memory device. There are several different types of external memory: magnetic tape, magnetic drums, and magnetic discs. The last type is further subdivided into hard discs and floppy discs.

Two types of magnetic tape system are in use: audio and digital. The audio tape system is slow (less than 1100 baud) and popular only among the low-cost end of the home computer market. Few, if any, professional or business systems will use audio tape. In these systems, the bits are represented by two tones, one for HIGH and another for LOW. In my own Digital Group system, for example, the HIGH condition is represented by 2125 Hz and LOW is represented by 2975 Hz. The decoding method is similar to those discussed earlier.

There are only a few advantages to audio tape systems. One is low cost. The tape system used is ordinary audio cassettes. In actual practice, any good quality audio cassette machine is usable, and the tapes are the so-called middle-quality. As a result, program entry and storage are made available to low-end machines. But audio cassettes are also full of problems. Even at slow data rates there are frequent misreads. At the 1100 baud rate of the Digital

Group TVC interface, read errors are far more common unless very good quality machines and tapes are used. Tape speed is also very critical in audio cassette systems. Recall from Chapter 14 that tone decoders tend to use filters to separate the High and LOW tones. If the speed is in error, then the tones received from the recorder will be outside of the passband of the filter (or alternatively, the phase-locked loop capture range). But you will see audio cassette interfaces in most home computers, where cost is a determining factor in design choices.

Digital tape uses special magnetic oxides that can be magnetized by passing a pulse current through the record head. The bit is represented by the magnetic state at the correct point on the tape. An example of digital magnetic tape machines is the Phi-Deck machine. Many users of digital cassette machines use ordinary audio-grade cassettes, but unlike audio cassette systems, must use only the highest grade of cassette. Some manufacturers offer certified computer-grade cassettes for these machines.

Large systems use reel-to-reel tape drives that are very high quality machines. In many of these machines, the tape is placed against the head more firmly by vacuum pressure. These machines are large drives and very costly, so they are found only in larger commercial computers.

An example of a magnetic drum memory is shown in Fig. 23-8. The drum is a cylinder whose surface is coated with a magnetic material similar to that used in magnetic tape (a ferric oxide). A read head is positioned so that it could move up and down in the vertical plane; then any point along that line can be read. The drum rotates, so by knowing where on the vertical plane and where on

Fig. 23-8. Magnetic drum memory.

the rotation cycle a bit is located, any given data can be found. In order to know where the drum is in the rotation cycle, some timing bits must be produced. These bits are usually along one edge of the drum and are read by a fixed head. Many timing bits are received each cycle, and they can be electronically counted. One additional bit, a synchronization bit, is needed. There will be only one sync bit per rotation. If this bit is used to reset the timing counter to zero once each revolution, then the state of the counter, as determined by the timing bits, will locate the position of the drum for the computer.

Discs are also magnetic devices. The hard disc is coated with ferric oxide and resembles an oversized phonograph record. Hard discs are used in minicomputers and larger computers. In smaller systems based on microcomputers, however, floppy discs are used. These are similar in size to the old 45 rpm phonograph records. Both 5-inch and 8-inch diameter models are available. They will not operate at the same speed as the hard discs and will not store nearly as much data, but they are a low-cost alternative to mass storage.

PLOTTERS/RECORDERS

One last class of device which we will consider is chart recorders and plotters. Most of these are analog instruments, but some are digitally operated. Analog readers are oscillographic, and will reproduce the waveform of the applied signal onto graph paper. An analog type recorder must be driven by an analog signal, usually a voltage. If we are to drive such an instrument with a computer, then we must first convert the binary output of the computer to a voltage. This is done by a digital-to-analog converter (DAC), as described in Chapter 19.

Analog recorders use either permanent magnet moving coil (PMMC) galvanometers that are much like D'Arsonval meter movements, or the X-Y servomechanism. But digital recorders use stepper motors to drive the pen (one each for vertical and horizontal). A stepper motor will advance only when a pulse is received. A constant level (either HIGH or LOW) will not cause the motor to advance; only a *pulse* will do the trick. When these devices are used, the stepper motor for each axis is connected to one bit of an output port or to a serial output port. These devices are analogous to *mechanical DACs*, with the analog output being pen position.

Appendices

Appendix A

Logic Level Detector Circuits

Fig. A-1. Logic level detector circuit.

Fig. A-2. Logic level detector circuit.

Fig. B-1. A four-channel oscilloscope switch.

Index

Index

421

Other Bestsellers From TAB

Other Bestsellers From TAB